KAVACH

Cab Signalling and Automatic Train Protection System for Digital Railways

Lalit Kumar Mansukhani, B. E. Hons
OSD/CoE/IRISET
IRSSE, NF-HAG

Chandra Kishore Prasad, B. Tech, M.B.A in IT
Sr. Professor in IT & KAVACH, IRISET
IRSSE, NF- HAG

Y Ramesh Babu, M. Tech
Sr. Professor in Signal, IRISET
IRSSE, SAG

Published in association with

इरिसेट IRISET

by

BSP BS Publications
An Imprint of **BSP Books Pvt. Ltd.**

KAVACH - Cab Signalling & Automatic Train Protection System for Digital Railways
By Indian Railways Institute of Signal Engineering & Telecommunications

Published by BS Publications in association with © Indian Railways Institute of Signal Engineering & Telecommunications

BS PUBLICATIONS
An Imprint of BSP Books Pvt., Ltd.
4-4-309/316, Giriraj Lane, Sultan Bazar,
Hyderabad - 500 095
Phone : 040 - 23445688
e-mail : info@bspbooks.net
www.bspbooks.net

INDIAN RAILWAYS INSTITUTE OF SIGNAL ENGINEERING & TELECOMMUNICATIONS
Secunderabad - 500 017

ISBN: 978-93-95038-35-5 (Hardback)

FOREWORD

Digitalization has revolutionized the way we live and work, and the railway industry is no exception. Digital technologies will play a critical role in improving efficiency, safety, and overall passenger experience in future transportation systems. The book, "Kavach: CAB Signalling and Automatic Train Protection System for Digital Railway," is a comprehensive guide to one of the most important digital systems in the railway industry - the Kavach Cab Signalling and Automatic Train Protection (ATP) system.

The Kavach system, developed by Indian Railways, is a state-of-the-art digital technology that enables real-time monitoring of train movements and provides advanced automatic train protection capabilities. This system is critical for ensuring the safe and efficient operation of trains in the digital railway era.

The book covers all the necessary topics that are required for the students to understand the technology behind Kavach. In particular, the book provides an in-depth examination Kavach architecture, design, and implementation. It covers various components of the system, such as on-board and trackside units, and the ways in which they interact to provide a safe and efficient railway system. The book has given a detailed overview of communications systems and network monitoring systems of Kavach. The book also explores the benefits of Kavach, including improved train control, increased capacity, and reduced risk of accidents.

I complement the authors who have done an excellent job of presenting the complex technical details of Kavach in a clear and concise manner, making it accessible to a wide range of readers, including railway professionals, students and researchers in the field of digital railway systems.

This book is an invaluable resource practicing railway engineers, managers, and policymakers. It is also a valuable resource for students and researchers studying digital railway systems. I believe that this book will be an invaluable resource for anyone who is interested in understanding the technology behind digital railway systems and how they are helping to shape the future of transportation.

I highly recommend this book to anyone with an interest in digital railway systems and the future of transportation.

Dr. Abhay Karandikar
Director,
Indian Institute of Technology Kanpur

FOREWORD

Train is the people's mode of transportation in India. Indian Railways cater to all classes, from the poor to the most luxurious. Railways provide affordable means for the middle and lower classes to travel to other parts of the country in search education, employment, and leisure. Matching opportunity with availability of people is critical to a nation's growth. As a student I used to travel 2245 kms from my hometown of Alwaye in Kerala to reach IIT at Kharagpur in West Bengal for education. Indian Railways, in a sense, sponsored my education partially! Faster and safer trains will continue to be the blocks to build an India of greater prosperity.

The digital revolution combined fast & cheap computing with fast & cheap communications. The digital revolution has transformed the world in ways we never imagined earlier. Train tickets were physical assets to be guarded dearly until recently; today they are disrupted into being a PNR number. Money became a number on a banking app and not wads of currency notes. The world has been made more efficient and cheaper using cutting-edge hardware, communication, and software technologies in recent decades. Rail travel is the safest mode of modern transportation. However, the societal demand for scale, quality, and safety can only be met using greater deployment of digital technology and information processing to improve efficiency, safety, and the overall passenger experience of Indian Railways.

This book, "Kavach: CAB Signalling and Automatic Train Protection System for Digital Railway," is a comprehensive guide to one of the most important digital systems in the railway industry - the Kavach Cab Signalling and Automatic Train Protection (ATP) system. Kavach is a bold and timely attempt to exploit advances in sensing, computing, and communication technologies to greatly enhance the safety and efficiency of trains. The boundary between hardware, software, sensor, communication, etc., is getting blurred by modern technologies. Kavach embodies an integrated approach as can be seen in its design. Kavach provides automatic train protection capabilities while enabling real-time monitoring of train movements

Systems like Kavach become the engineering marvels of tomorrow. It is important that such real-life systems be included in the engineering curricula to impart practical insights into the theory taught in classes. This book is designed to be used as a textbook for a course approved by AICTE. It covers the required topics to understand the technology behind Kavach and provides an in-depth examination of its architecture, design, and implementation. It covers the various components of the system, such as the on-board and trackside units, and the ways in which they interact to provide a safe and efficient railway system. It also explores the benefits of Kavach, including improved train control, increased capacity, and reduced risk of accidents.

The authors have done a commendable job of presenting complex technical details in a clear and concise manner, making it accessible to a wide range of readers, including railway professionals, students, and researchers in the field of digital railway systems. I believe this book will be an invaluable resource for anyone who is interested in understanding the technology behind digital railway systems and how they are helping to shape the future of transportation. I highly recommend this book to anyone with an interest in digital railway systems and the future of transportation.

Dr. P J Narayanan
Director, IIIT Hyderabad

PREFACE

Over the past few decades, the world has undergone a significant shift towards digitalization, and the railway industry has not been left behind. The Indian Railways, one of the largest railway networks globally, has taken the lead in adopting digital technologies to improve train safety and efficiency.

"Kavach – Cab Signalling based Automatic Train Protection System for Digital Railway" is a comprehensive guide that explains the indigenously developed technology for train protection and safety jointly created by the Indian Railways and Indian eco-system. This book covers the key concepts and technical aspects of the Kavach system, which is set to revolutionize train safety and operations in India.

The Kavach system offers monitoring and control of train movement, constant supervision of speed, and automatic braking in case of any potential danger. It comes with train and track-side sensors, communication systems, track-side and on-board vital processors compliant to SIL 4 architecture, and algorithms that quickly detect and respond to any threat to train safety.

The technology behind Automatic Train Protection Systems can be classified into two categories based upon mechanism of exchange of information between trackside and train units - Radio based and non-radio based. Kavach- Cab signalling & Automatic Train Protection system is a Radio based system designed to enhance train traffic throughput and adaptable to newer functionality to meet the future roadmap of Indian Railways.

It is a remarkable feat for a country to have their own indigenously developed radio-based ATP (Automatic Train Protection) with cab-signalling system for their mainline railway. In fact, only a handful of nations in the world have been able to achieve this level of technological advancement. This accomplishment reflects the country's commitment to innovation and infrastructure development, as well as their ability to compete on a global scale in the transportation industry

The AICTE has approved a course on Kavach, an indigenous ATP system, for Engineering UG programs as an open elective. This book targets Bachelor of Technology students, railway professionals, and researchers interested in digital train control and protection technology. It provides an in-depth understanding of the Kavach system's concepts, design, operation, communication systems, and implementation.

The Kavach system demonstrates the Indian Railways' commitment to innovation and technological advancement. We believe that this book will be an excellent resource for anyone interested in learning more about this ground breaking technology.

We extend our sincere thanks to all the contributors and authors who helped create this book. Their expertise and knowledge were crucial in making this book a comprehensive guide to the Kavach system.

We hope that this book will inspire future generations of railway professionals to continue pushing the boundaries of technology and innovation to make our railways safer, more efficient, and more sustainable.

-Authors

ACKNOWLEDGEMENT

We would like to extend our deepest gratitude to the head of our institute, Shri Sudhir Kumar, Director General for his unwavering support and guidance for writing the book. His encouragement and belief in our work has been invaluable, and we are grateful for his contributions to the success of this book.

We would also like to extend our thanks to Shri Abhay Karandikar, Director, IIT Kanpur and Shri P J Narayanan Director IIIT, Hyderabad, the writer of the foreword, for their insightful comments and invaluable feedback. Their perspective has helped us to refine our ideas and shape the final product.

We would also like to express our appreciation to Shri G. K. Dwivedy, Principal CSTE and Shri V. N. M. Rao CSTE Project from South Central Railway for providing valuable insights that have helped shape the content of this book.

Finally, we would like to thank the book reviewers, Smt L. Nirmala Devi, HOD Electronics Engineering, OUCE, Hyderabad and Shri Uma Shankar Sahu, Professor NITTTR, MoHRD, Hyderabad for their time and effort in critically evaluating our manuscript. Their comments have helped us to improve the book and make it more accessible to our readers.

We could not have done this without the help and support of these individuals and the teaching faculty of IRISET, and we are eternally grateful for their contributions to this book.

-Authors

ABOUT THE AUTHORS

Lalit Kumar Mansukhani, IRSSE, OSD, CoE for Modern Signalling, IRISET

Lalit Kumar Mansukhani is All-India topper of his IRSSE batch and holds more than 30 years of experience in Railway Industry including 4 years of the UK Railway Signalling Design House. He has also served in ISRO for 4 years. He is recipient of several awards and accolades. He is regarded as the prime architect of Cab signalling based IR-ATP Kavach and its radio communication backbone, MComm. As Director/Signal/RDSO, he pioneered the core R&D and coordinated the field trials in association with three Indian firms in 265 kms section which included Train Traffic Throughput Enhancement. His name appears as an inventor in the Kavach's Patent application. He made inhouse presentation of Kavach to the Hon'ble Prime Minister of India in 2016. Currently, he is heading Indian Railways' Centre of Excellence for Modern Signalling-Kavach at IRISET as Officer on Special Duty.

Chandra Kishore Prasad, IRSSE, Sr Professor (IT & Kavach) IRISET

He has served in various positions on Indian Railways in areas encompassing Operations and Maintenance, Projects, OFC, Microwave Transmission & Switching System, Planning, Performance Management, Training and Disaster Management and RailTel in his 30 years career in Rail Transportation. He served as Executive Director, on deputation to RailTel Corporation of India Ltd, in-charge of Southern Region and was instrumental in building of Tier III Uptime Facility Certified Data Centre at Hyderabad, Design to Operation of Cloud services based on Software Defined Data Centre architecture and Green Edge Datacentres at 5 cities. He is from Indian Railway Service of Signal Engineers- 1991 batch.

He has received many awards and accolades for his Digital Transformation initiatives notable being IDC, Data Quest, Indian Express, and Business Transformation Award from Economic Times & Hindustan Times media. Currently he is engaged in training development & capacity building for indigenously developed ATP-Kavach, Proof of Concept for LTE and Working group on 5G. He is closely working with AICTE & MoSDE for skill development in Signalling & Kavach for various technical roles in addition to the Digital Transformation initiatives at IRISET.

Y. Ramesh Babu, IRSSE, Sr Professor Signal, IRISET

He belongs to Indian Railway Service of Signal Engineers- 1998 batch. He was associated with projects of signalling installations and commissioning including Railway Electrification and Electronic Interlocking. At present, he is working as Senior Professor/Signal in Indian Railways Institute of Signal Engineering and Telecommunications (IRISET). He oversees the simulation and laboratory set up of KAVACH available for trainees at IRISET.

ABBREVIATIONS

Abbreviation	Description
ABS	Automatic Block Signalling
AC	Alternating Current
ACK	Acknowledge
ASCR	Advanced Starter Signal Control Relay
ATP	Automatic Train Protection
ATS	Actual Toe of Switch
BER	Bit Error rate
BIU	Brake Interface Unit
BSLB	Block Section Limit Board
BTM	Balise Transmission Module
CD	Compact Disc
CE	European Conformity
CRC	Cyclic Redundancy Check
CPRI	Central Power Research Institute
CPU	Central Processing Unit
CCTV	Closed Circuit Television
CENELAC	French Acrynom- Comite' Europe'een deNormalisation E'lectrotechincal Standardizatio
CTS	Clear To Send
CUG	Closed User group
dB	Decibel
dBi	Decibel relative to isotrope
dBm	Decibel Mill watts
DC	Direct Current
DECR	Green Lamp Checking Relay
DEMU	Diesel-Electric Multiple Unit
DG	Diesel Generator
DMI	Driver Machine Interface
DMU	Diesel Multiple Unit

DWC	Double Walled Corrugated
E1	2048kbps Data Channel
EC	Emergency Communication
ECR	Lamp Checking Relay
EI	Electronic Interlocking
EMI	Electromagnetic Interference
EMU	Electrical Multiple Unit
EOA	End of Authority
EPC	Evolved Packet Core, Engineering Procurement & Construction
FAT	Factory Acceptance Test
FCC	Federal Communications Commission
FDMA	Frequency Division Multiple Access
FRMCS	Future Railway Mobile Communication System
FS	Full Supervision
FSK	Frequency Shift Keying
FSS	First Stop Signal
G	Gate
GI	Galvanized Iron
GNSS	Global Navigation Satellite System
GPS	Global Positioning System
GPRS	General Packet Radio Service
GSM	Global System for Mobile Communication
HDPE	High Density Poly Ethylene
HECR	Yellow Lamp Checking Relay
HHECR	Double Yellow Lamp Checking Relay
HQ	Headquarters
IBS	Intermediate Block Signalling
IC	International Company
IPS	Integrated Power Supply
IRS	Indian Railway Standards
IS	Isolation, International Standards
JE	Junior Engineer

JPO	Joint Procedure Order
KA	Authentication Key
KMPH	Kilometer Per Hour
KMS	Key Management System
KS	Session Key
LC	Level Crossing
LCD	Liquid Crystal Display
LED	Light Emitting Diode
LKAVACH	Loco KAVACH
LP	Loco Pilot
LS	Limited Supervision
LSS	Last Stop Signal
LTE	Long Term Evolution
LXPR	Level Crossing Gate Proving Relay
MA	Movement Authority
MAC	Message Authentication Code
MB	Mega Byte

CONTENTS

CHAPTER 1: KAVACH - AN OVERVIEW

CHAPTER

1 KAVACH – An Overview

1.0 Introduction

KAVACH is an indigenously developed Automatic Train Protection (ATP) System meant to provide protection to the trains against Signal Passing at Danger (SPAD), excessive speed and collisions. KAVACH provides continuous update of Movement Authority (distance up to which the train is permitted to travel without danger). Hence during unsafe situations when brake application is necessitated, and the Crew has either failed to do so, or is not in position to do so, automatic brake application shall take place. KAVACH has additional features to display information like speed, location, distance to signal ahead, Signal aspects etc. in Locomotive cab and generation of Auto and Manual SOS messages (Distress messages) from Locomotive as well as from the Station unit in case of emergency situation. The communication between Stationary KAVACH and Loco KAVACH units shall be Safety Integrity Level 4 (SIL 4) certified as per the CENELEC standards for railways. Communication between two or more Loco KAVACH systems, Non-Signalling based additional collision protection (i.e. Head-on, Rear-end & Side Collisions) and Manual SoS are non-SIL based (i.e. not failsafe).

Spot or Intermittent update type system affects Line Capacity adversely due to inherent limitation of design. Continuous Update type of system like KAVACH is Line Capacity friendly by design. Continuous Update type European system has been using commercially off-the shelf mobile communication such as GSM-R for long. KAVACH uses innovative radio communication system designed by RDSO to achieve continuous communication between the Train and Track-side systems.The radio communication system operates in UHF band.

Safety Integrity Level (SIL): The concept of Safety Integrity Level (SIL) is a direct result of the IEC 61508 standard, which is not specific to railways. For the railway industry, CENELEC (European Committee for Electrotechnical Standardization) has developed the standards EN 50126 (The Specification and Demonstration of

Reliability, Availability, Maintainability and Safety (RAMS)), EN 50128 (Communication, signalling and processing systems - Software for railway control and protection systems), and EN 50129 (Communication, signalling and processing systems - Safety related electronic systems for signalling) which have been derived from the IEC 61508 standard to meet the specific requirements for the design, development, and assessment of safety-related electronic systems used in the railway industry.

SILs are assigned based on the level of risk associated with a particular system. The risk is determined by the likelihood and severity of a hazardous event occurring. The likelihood is calculated based on the probability of a component or system failure, while the severity is assessed based on the potential consequences of a hazardous event.

There are four Safety Integrity Levels, labelled from SIL 1 to SIL 4: The latter has the highest level of safety integrity and SIL 1 the lowest. The costs increase significantly to reach higher levels of SIL, however the cost of not implementing the appropriate SIL far exceeds the cost of the overall implementation.

The standard associates Safety Integrity Levels with numerical probabilities of hazardous failures for systems that operate continuously like Kavach (Table 1.1)

Table 1.1 SIL for Continuous Demand Mode

Safety Integrity Level	Probability of dangerous failure per hour (Continuous demand mode of operation)
SIL 1	$\geq 10^{-6}$ to $< 10^{-5}$
SIL 2	$\geq 10^{-7}$ to $< 10^{-6}$
SIL 3	$\geq 10^{-8}$ to $< 10^{-7}$
SIL 4	$\geq 10^{-9}$ to $< 10^{-8}$

As the SIL level increases, the Dangerous Failure Rate must decrease accordingly. Therefore, higher SIL levels require a higher level of reliability and safety, and more rigorous design and testing procedures.

1.1 KAVACH Features

1. Prevention of Signal Passing at Danger (SPAD)
2. Cab-Signaling, Loop Line Speed Control
3. Prevention of Over speed: Section Speed, Train Speed, Permanent Speed Restriction (PSR)
4. Protection of Roll back and Reverse movements
5. Prevention of Side-collision in block section
6. Prevention of Head-on & Rear end collisions
7. LC Gate Automatic Warning

8. Save Our Souls (SOS) Messages
9. Computation of Train Length
10. Shunt Limits Validation
11. Centralized live monitoring of Train movements in Networking Monitoting System (NMS).

1.2 System and Sub Systems

The brief overview of functioning of KAVACH is as given below:

- The trackside sub-system of KAVACH consists of Radio Frequency Identification (RFID) tags fitted on track in station section and block section for giving Trackside information to Loco KAVACH unit installed in the locomotive. Portions of track including berthing tracks, point and block sections are assigned unique IDs called Track Identification Number (TIN).
- The system also consists of Stationary KAVACH unit installed at Station with radio tower to communicate with locomotives in the jurisdiction of Stationary KAVACH which commence on approach of its first Signal. Stationary KAVACH is interfaced with station interlocking to acquire real-time dynamic information related with signalling such as various signal aspects. Route information of all the signals monitored by a specific stationary KAVACH unit is configured on the basis of KAVACH Control Table (excluding shunt signals and overlaps). Stationary KAVACH unit gets real-time information regarding Locations, Speed etc of various trains in its jurisdiction through UHF Radio Communication.
- Separate *Stationary KAVACH unit* is provided at Mid–Section interlocked Level Crossing Gate and Intermediate Block Signalling (IBS) locations if they do not come within the coverage of station radio tower.
- Remote Interface Unit (RIU) shall be used where remote signalling functions are required to be fetched to a nearby Stationary KAVACH unit for example from end cabins/ distributed interlockings or LC gate/IB coming within the radio coverage of station tower.
- The *onboard Loco KAVACH*unit installed in the locomotive determines the location of train by reading pre-programmed *RFID Tag data* with the help of *RFID reader.*
- By using the cryptographic technique malicious attacks will be eliminated. Only Authenticated Systems communicate *with each other* to ensure the safety of the system.
- Key Management System (KMS) is developed to check for authentication and distribution of secret keys.

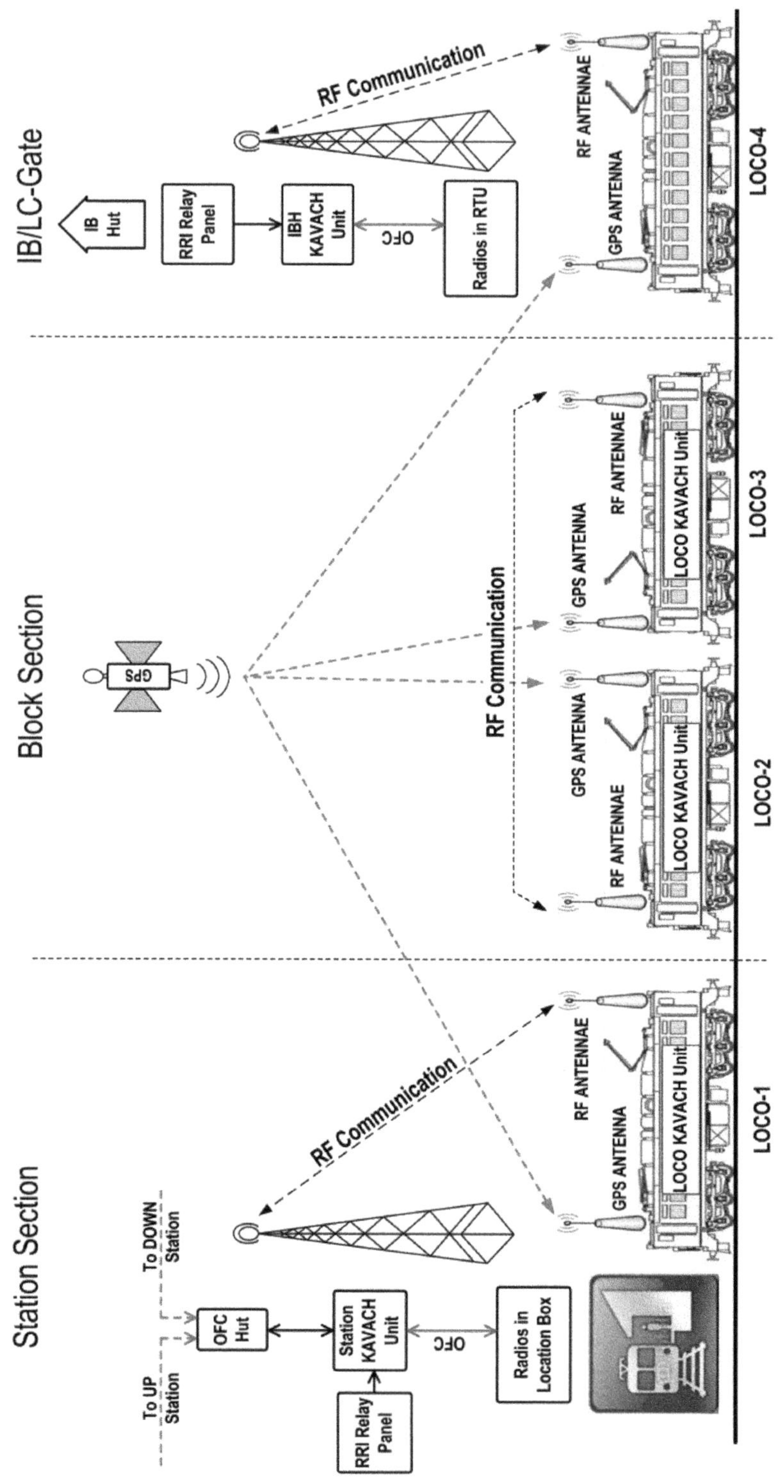

Fig. 1.1 KAVACH Automatic Train Protection

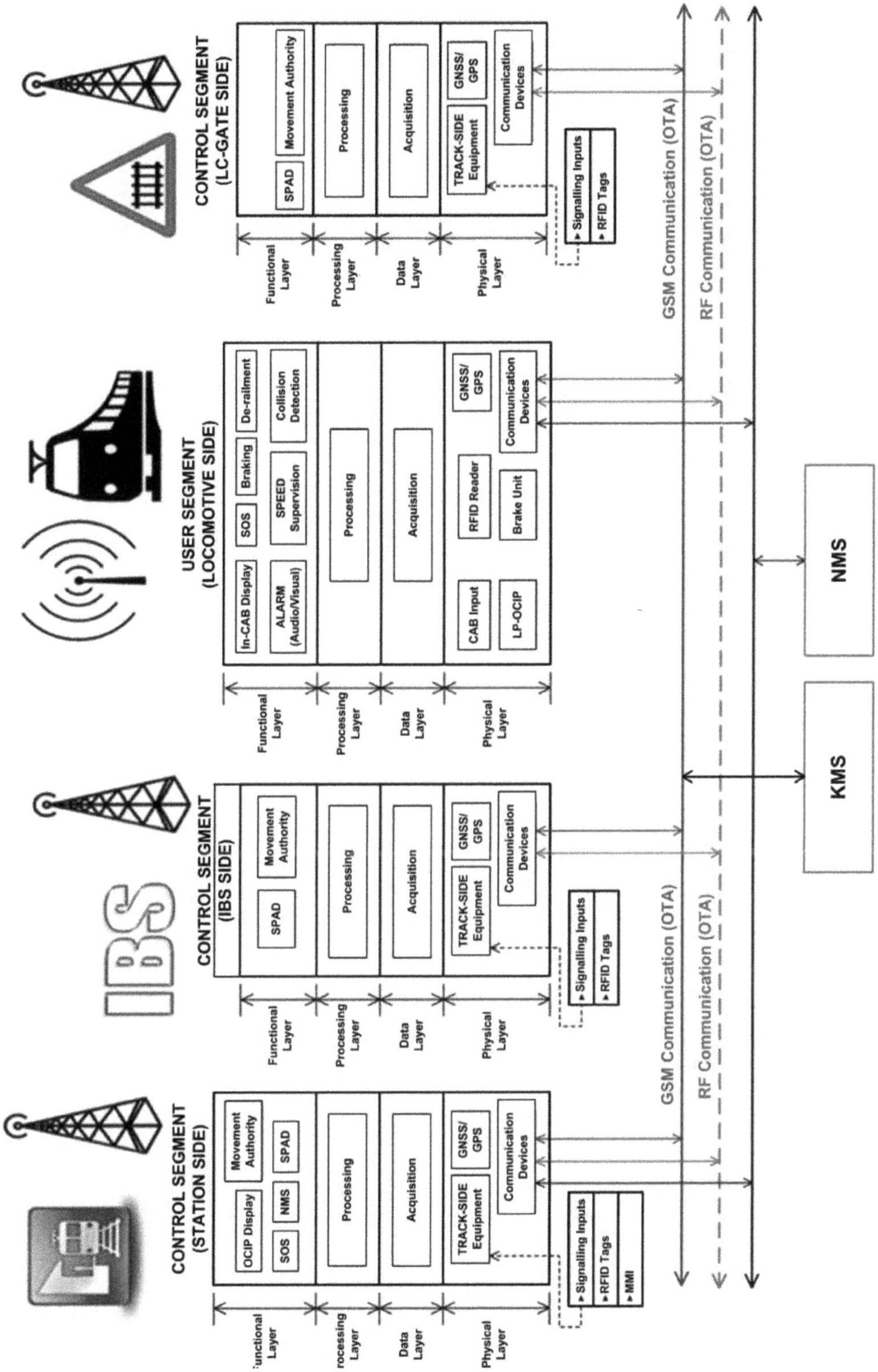

Fig. 1.2 KAVACHProcess flow

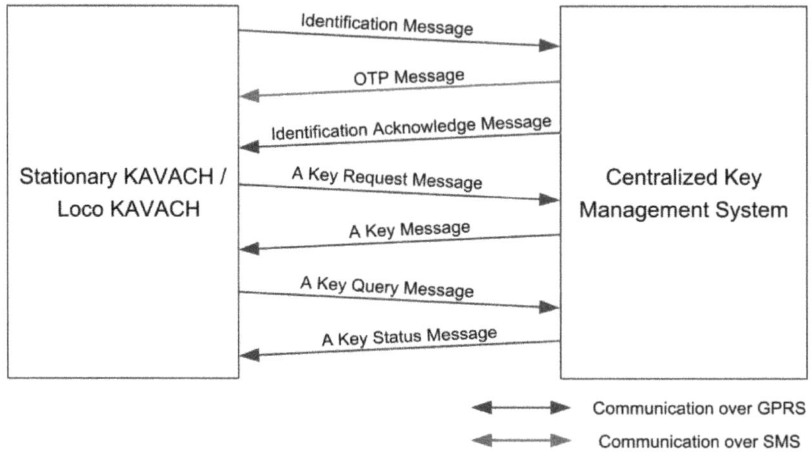

Fig. 1.3 Process flow for authentication keys transmission

- Loco KAVACH unit sets its absolute location (Approaching signal distance from the train position) and Track Identification Number (TIN) as undefined (zero) before determining the direction.
- The direction of movement of train shall be determined, when Loco/Train has passed two RFID tags sequentially with Absolute location.

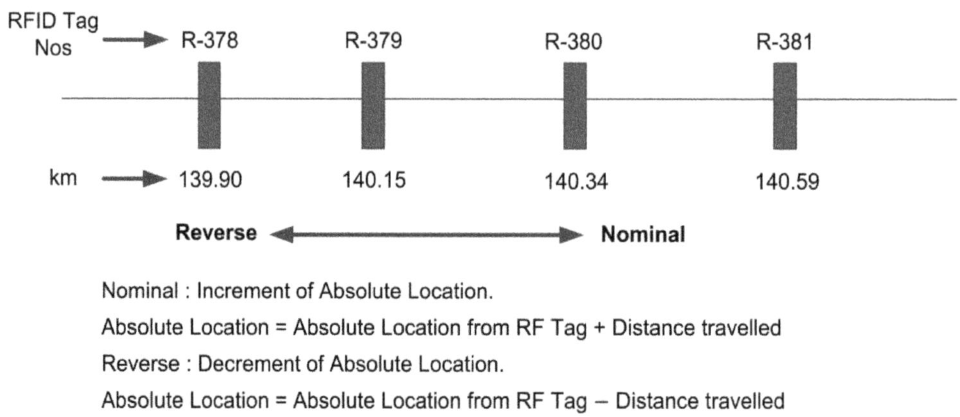

Fig. 1.4 RFID Normal and Reverse direction

- Loco KAVACH unit calculates the location of the train between two RFID tags dynamically based on the distance travelled from last RFID tag through speed sensing arrangement provided on Locomotive.
- On passing through the RFID Tag, Loco KAVACH unit transmits the location and direction of the train to the Stationary KAVACH unit through UHF radio antenna provided in locomotive.

- The Stationary KAVACH unit shall use the direction of movement of Loco/Train, to find approaching signal of the Loco/ Train.
- Stationary KAVACH unit shall then calculate the movement authority based on the signal aspect or/and track circuit status or/and route locking status, point position and the status of the berthing track circuit.
- Stationary KAVACH unit shall then transmit the Movement Authority to the Loco KAVACH in its jurisdiction in station area.

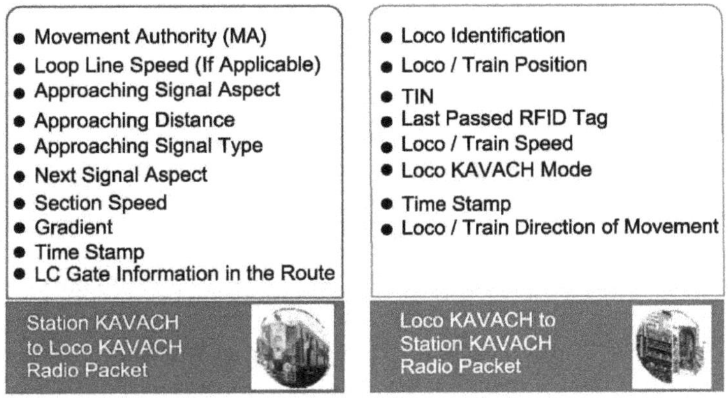

Fig. 1.5 Radio packet between Loco and Stationary KAVACH

- The length of the movement authority is decided based on the signal aspect of the approaching Stop Signal.
- The Loco unit shall make speed profile/ brake curve for different situations based on movement authority, speed restriction and other information as received from Trackside sub-system.
- The Loco KAVACH unit shall display the train speed, the permitted speed, the target distance and the target speed to the loco pilot through a *Driver Machine Interface (DMI)*.
- If a signal on approach is Red (Danger), the Stationary KAVACH unit shall transmit this information to the Loco KAVACH and reduce the movement authority to zero.
- If loco pilot fails to stop the train, automatic application of brakes shall take place, thus preventing *Signal Passing at Danger (SPAD)*.
- In case of any conflict between signal aspect, point position, berthing track section, signal aspect sequence and TIN, the Stationary KAVACH unit shall transmit most restrictive aspect of that signal and shall reduce the movement authority accordingly. In this way train collisions are prevented in the station section.
- In case of block section if two trains are detected by Stationary KAVACH to be moving towards each other on same TIN, the SoS command would be generated by Stationary KAVACH for both the trains.

- On reception of such Loco specific SoS from Stationary KAVACH Unit, the Trains would be stopped through automatic application of brakes. There is also provision for broadcasting SoS message from Loco KAVACH to other Loco KAVACH in case of emergencies.
- Communication technique used for transfer of information between Stationary and Locomotive units in station area is *Full Duplex UHF Radio Communication* through *Multiple Access Time Division Multiple Access (TDMA)/ Frequency Division Multiple (FDMA) Access scheme.* A specific frequency pair is allotted to a station for communication between Stationary KAVACH and Loco under its jurisdiction.
- Loco KAVACH units can also communicate with other Loco KAVACH units in block section, in station area and in emergency situations (SoS, head-on, rear-end collisions) using a fixed frequency (f0) in its designated time slot.
- For centralized monitoring of KAVACH equipped Trains and Stations within the network, Network Monitoring System (NMS) with a central server in divisional office shall be provided over OFC Network. Transmission of exceptional fault/critical messages from Stationary KAVACH as well as Loco KAVACH to NMS is done through respective GSM interfaces available to them. Trouble shooting of error events, off line simulation, real time monitoring of KAVACH loco etc. are done through NMS.
- In the KAVACH System Radio Communication shall use cryptographic techniques to transfer messages between Loco KAVACH and Stationary KAVACH units. For secured communication, Authentication keys are received by Stationary KAVACH and Loco KAVACH using Global system for Mobile Communications (GSM)/General Packet Radio Service (GPRS) communication through a Key Management System (KMS). Real Time Clocks (RTC) of all the KAVACH systems are synchronized with Global Positioning system (GPS)/Global Navigation Satellite System (GNSS).

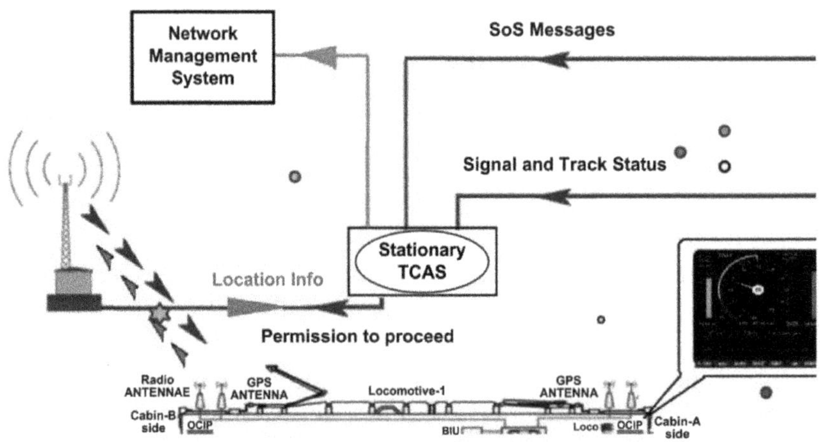

Fig. 1.6 KAVACH Functioning Schematic

1.3 KAVACH Requirement and Implementation

Requirement	Mechanism in KAVACH
Direction of trains	Comparing Absolute location of two RFID tags passed by train.
Location of Trains	Distance traversed beyond a RFID tag on Track Sleeper (Rail-road Tie) through speed sensing arrangement (Tachometer)
Extraction of dynamic Signalling Information by stationary KAVACH of station/ IB/ LC gate vital computer.	Interfacing to station interlocking (PI/RRI/EI)
Transfer of Signalling related information from Station KAVACH to Train	Radio Communication between Stationary unit & Train units through dynamic TDMA on a specific frequency pair in station area. Stationary units are allocated timeslots according to Topography and their size. Mobile units i.e. trains are assigned slots dynamically. This provides efficient utilization of channels.
Loco to Loco message broadcast	In block section, in station area and in emergency situations (SoS, head-on, rear-end collisions) using a fixed frequency (f0) in its designated time slot.
Prevention of over speed and SPAD	By reducing the movement authority based on the aspect of approaching signal.
Prevention of collisions between two trains	Through conflict between signal aspect, point position, berthing track section, signal aspect sequence and TIN in station area and through TIN conflict in block section.
Centralized monitoring of KAVACH equipped trains and stations	Through Network Monitoring System (NMS)
Security of radio communication between Stationary KAVACH and Loco KAVACH	Using GSM/GPRS communication techniques through a Key Management System (KMS)
Real Time Clock (RTC) synchronization	Through Global Positioning system (GPS)/ Global Navigation Satellite System (GNSS).

1.4 Data Flow between KAVACH Sub Systems

Information received by Stationary KAVACH unit from Loco KAVACH unit

- Direction of train movement
- Location of train
- Emergency messages

Information received by Stationary KAVACH unit from interlocking

- Approaching Signal aspect (ECRs)
- Berthing track circuit status (TPRs)
- Point status (NWKRs/RWKRs)
- Status of block instrument - Line Closed condition

1.5 Data sent by Stationary KAVACH to Loco KAVACH

- Aspect of the approaching signal on route.
- Approaching signal distance from the train position (absolute location).
- Approaching signal identity.
- Next signal aspect and its distance in the territory of same stationary KAVACH unit, if signal on approach is OFF.
- Movement authority (the distance for which the train is authorized to travel).
- Static Speed Profile.
- Temporary Speed Restrictions

1.6 Determination of Direction

- The direction of movement of train shall be determined through RFID Tags.
- There shall be three types of direction of movements, one for train such as forward or reverse, second for traffic such as UP or DN and the other for movement of direction such as Nominal or Reverse.
- The direction shall be derived when Loco/Train has passed two RFID tags with Absolute location (except Adjustment tag).
- If Absolute location value is incrementing, it shall be treated as Nominal direction. If Absolute location is decrementing, it shall be treated as Reverse direction.
- The direction of movement of train and TIN shall be used for determining whether two trains are approaching or one following the other or going away from each other.
- The Stationary KAVACH unit shall use the direction of movement of Loco/Train, to find approaching signal of the Loco/Train.

1.7 New Train Formation

A train shall be considered as a new formed train by Loco KAVACH unit under one or more of the following conditions:

- When Loco KAVACH unit has been switched on or restarted
- When Loco KAVACH unit has come out of Non-Leading/Shunt mode/System failure mode/Isolation mode.
- When driving cab/desk is changed except when Loco KAVACH is in Shunt mode.

1.8 Train Length Assisgnment

- Every stationary KAVACH unit shall monitor the status of track section identified for measurement of train length. (Only applicable to Station KAVACH units).
- Based on the time of occupation and clear status of these track sections, communicated by Stationary KAVACH unit, Loco KAVACH unit shall calculate its train length.
- Two track circuit (say AT & BT in sequence in the traffic direction of train movement) at the entry to block section shall be identified at each station for train length measurement. The track circuits identified shall be such that all the trains entering into a block section pass over these track circuits.
- The status of these track circuits shall be taken as input to stationary unit.
- Stationary unit shall communicate the time offset from frame cycle reference for 'BT occupied' and 'AT cleared' to concerned Loco unit, which shall be used by Loco unit for precise location for train length calculation.
- In case of overlapping territories, the train length measurement information shall be passed on by taking over station.

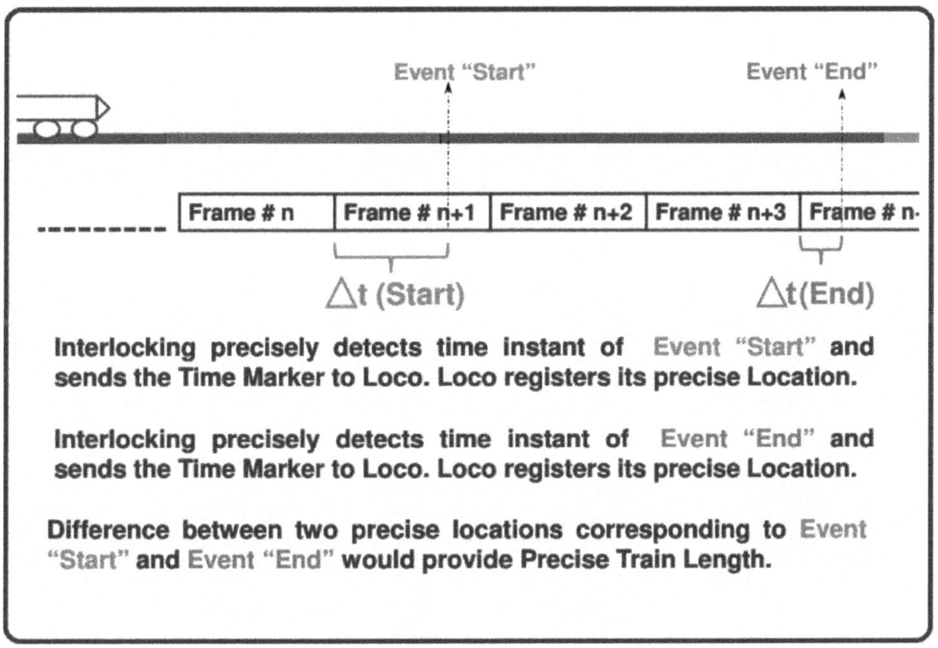

Fig. 1.7 Train length Measurement

1.9 Train Location

- The Loco KAVACH unit shall determine the location of the train with the help of RFID tag data and Speed sensor output.

- Loco KAVACH unit shall transmit the location of the train to the Stationary KAVACH unit every 2 seconds in Full Supervision mode.

1.10 Visual and Audio Warnings on the DMI

- As long as the current speed is less than or equal to permitted speed indicated on DMI, in Full Supervision mode, the Loco KAVACH Unit shall neither generate warning for brake application nor apply the brakes.
- Visual and audio warnings about expected brake intervention by Loco KAVACH unit shall be given to the Loco pilot to enable the loco pilot to react and avoid intervention.

1.11 Supervision of Movement Authorities and Speed Limits

- The Loco KAVACH unit shall supervise the end of movement authority, if this information is available onboard.
- A train shall be supervised to its static and dynamic train speed profiles.
- If the train speed exceeds the permitted speed by 2 kmph (configurable), warning for over-speed would be generated.
- If the train exceeds the permitted speed by 5 kmph (configurable), the Loco KAVACH unit shall execute a brake intervention along with warning until the actual speed is not more than permitted speed.

1.12 Operational Modes in Loco KAVACH

The KAVACH loco equipment shall be capable of supervising the following operational Modes:

S.No.	Modes	Responsibilities
1	Standby Mode (SB) 	1. This Mode is default mode. At Power-On or If No CAB is selected KAVACH comes into this Mode. 2. In This Mode Train is Supervised for Standstill. Any movement if detected it applies EB. 3. If any CAB is occupied, Based on New train Status It performs Brake system health test. 4. On test success it Suggests SR or Shunt Mode to Loco Pilot. 5. Radio Transmission is Active if it is in KAVACH Area
2	Staff Responsible Mode (SR) 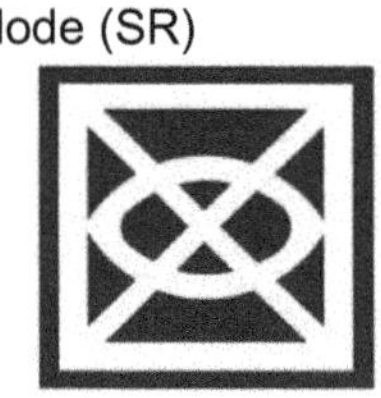	1. In KAVACH territory Transmits Radio packet. 2. Supervises Loco Ceiling Speed 3. It Supervise Roll away & Reverse Movement protection

Contd...

S.No.	Modes	Responsibilities
3	Limited Supervision Mode (LS) 	1. In KAVACH territory Transmits Radio packet. 2. Supervises Loco Ceiling Speed 3. If Track Data is Valid & Traffic Direction is Known It Supervises Section Speed (PSR), TSR (future use) & Collision Targets 4. It Performs Brake system health test 5. It obeys General SoS from Station if Loco is within 3km Radius 6. It Supervise Roll away & Reverse Movement protection. 7. If Radio communication is good supervises MA & Turnouts
4	Full Supervision Mode (FS) 	In this mode in addition LS mode 1. It Supervises MA, Turnout speeds received from Linked station 2. It performs Train Length Measurement 3. It Obeys Loco specific SoS from Linked Station
5	Override Mode (OV) 	1. Supervises OV Mode Speed 2. If MA is extended or after override timeout (Default: 240s), Section Speed is known & Traffic direction Known it will go to Full Supervision Mode 3. It Obeys SoS from Stationary KAVACH 4. Radio Transmission is Active if it is in KAVACH Area. 5. It can be entered only When MA
6	On-sight Mode (OS) 	1. It Obeys SoS from Stationary KAVACH 2. It Supervises On-sight Mode Speed (Configurable) 3. Radio Transmission is Active if it is in KAVACH Area
7	Trip Mode (TR) 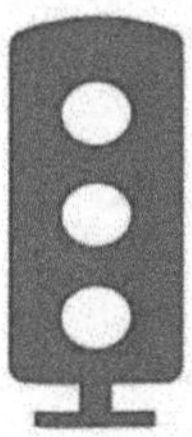	1. It enters in to this mode if (MA + 30m is crossed) or (Signal foot tag crossed after MA 2. Loco Pilot has to Acknowledge by pressing PTRIP soft key on DMI. Then it releases EB. 3. Stationary KAVACH shall send SoS to the Loco KAVACH unit until the trip mode is acknowledged. 4. Radio Transmission is Active if it is in KAVACH Area.
8	Post Trip Mode (PT) 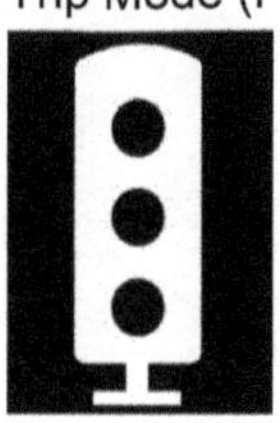	1. Supervises PT Mode Speed 2. If valid MA is received, Section Speed is known & Traffic direction Known it will go to Full Supervision Mode 3. It Obeys SoS from Stationary KAVACH 4. Radio Transmission is Active if it is in KAVACH Area

Contd...

S.No.	Modes	Responsibilities
9	Reverse Mode (RM) 	1. Supervises Reverse Mode Speed (Configurable), Distance (configurable) and Timeout (configurable) 2. It Obeys SoS from Stationary KAVACH 3. Radio Transmission is Active if it is in KAVACH Area
10	Shunt Mode (SH) 	1. Supervises Shunt Mode Speed (Configurable) 2. KAVACH doesn't do Brakes test if Loco pilot Changes AB.(After exiting Shunt mode It performs) 3. It Obeys SoS from Stationary KAVACH 4. Radio Transmission is Active if it is in KAVACH Area
11	System Failure Mode (SF) 	It Applies EB Continuously. Radio Transmission is Active if it is in KAVACH Area
12	Isolation Mode (IS)	It Bypasses Brakes. Loco KAVACH Unit shall transmit an Loco-to-Stationary Radio Packet at a periodicity of not less than 02 minute on encountering a KAVACH territory tag (which excludes LC Gate Tag) in one of the randomly selected access timeslots merely to indicate the onboard mode to NMS through Stationary KAVACH Unit

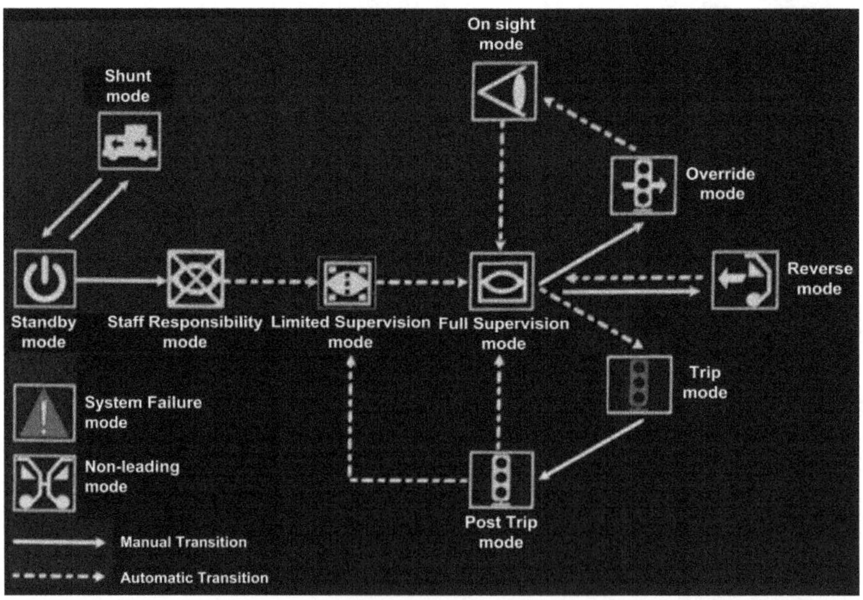

Fig. 1.8 Operational Modes flow chart

1.13 Static Speed Profile (SSP)

The Static Speed Profile (SSP) is a description of the fixed speed restrictions for a part of track sent from trackside to train. The SSP addresses the maximum permitted speed at any location. The static speed profile is a parameter that gives; the maximum permitted speed, that can be reached on the track based on physical characteristics of the track (curves, restriction to pass a point). The SSP is one of the data that allow the train to manage the supervision (Full Supervision, On Sight); for this the train will have to know the SSP throughout the Movement Authority.

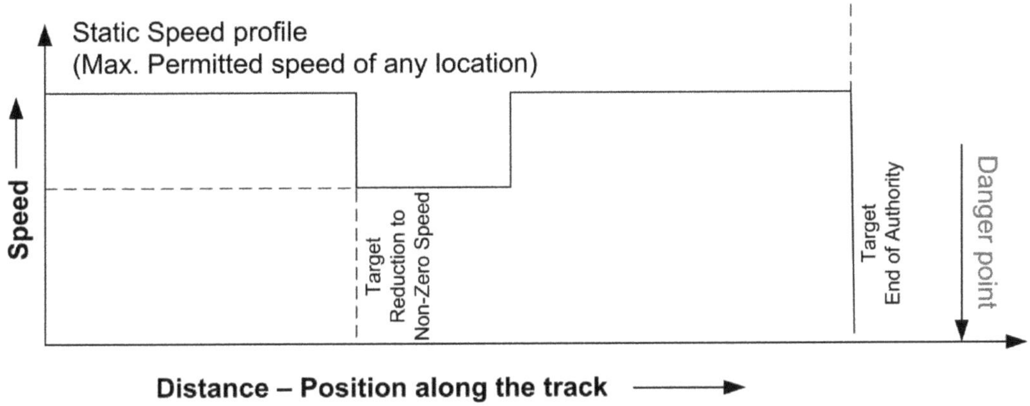

Fig. 1.9 Static Speed Profile (Distance-Speed graph) of a train

1.14 Dynamic Speed Profile (DSP)

Dynamic Speed Profile is the speed-distance curve which a train shall follow without violating the static train speed profile till the end of movement authority. This curve depends on the braking characteristics of the train and the train length. The dynamic speed profile considers the possible acceleration or deceleration curve of the movement of the train.

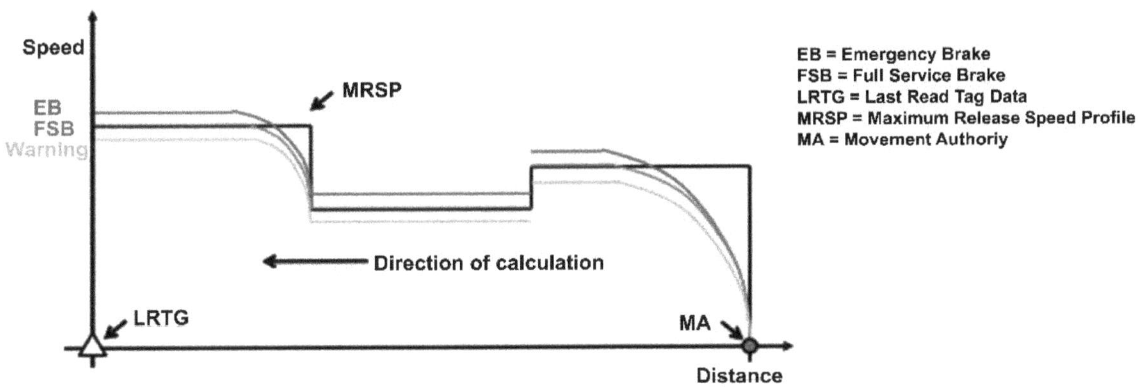

Fig. 1.10 Dynamic Speed Profile (Distance- Speed graph) of a train

1.15 Movement Authority

- Movement Authority (MA) is the distance upto which the train is permitted to travel without danger.
- The length of the movement authority is decided based on the signal aspect of the approaching Stop Signal.
- Aspect control chart is used to define the length of Movement of authority.
- In case of permissive signals, where the inputs for signal indications are available, the ECR shall be used for the purpose of displaying signal aspect. However, movement authority shall be decided based on the signal aspect of the approaching Stop Signal.
- In case of permissive signals, where the inputs for signal indications are not available, the signal aspect and movement authority shall be derived based on the signal aspect of approaching stop signal.
- The Movement Authority for the last signal of stationary KAVACH shall be the physical distance between the last signal of stationary KAVACH and the foot of next approaching Stop Signal. The Movement Authority shall be specified in meters. This movement Authority shall be used for train entering the block section.
- Stationary KAVACH unit shall calculate the movement authority based on the signal aspect or/and track circuit status or/and route locking status, point position, status of the berthing track circuit and status of the block instrument line closed condition. Stationary KAVACH unit then shall transmit the Movement Authority to the Loco KAVACH in its jurisdiction. The Movement Authority transmitted shall be the distance of End of Authority from actual Absolute Position of the train.
- For adapting KAVACH to an Auto section, it is necessary to communicate the signal aspects or/ and Track occupancy status to Stationary KAVACH, which then determines movement authority and communicates the same to the Loco KAVACH on radio.
- In the case of single line working, KAVACH shall extend Movement Authority after ensuring the establishment of direction of traffic and all stop signals (if available) against the established direction shall be at ON.

1.16 Sub Systems of KAVACH

The KAVACH (Indian Railway Train Protection System) broadly comprises of following components:

1.16.1 Track side Sub-systems

The Trackside subsystem shall be composed of

(i) RFID tag
(ii) Stationary KAVACH Unit
(iii) Tower and Antennae

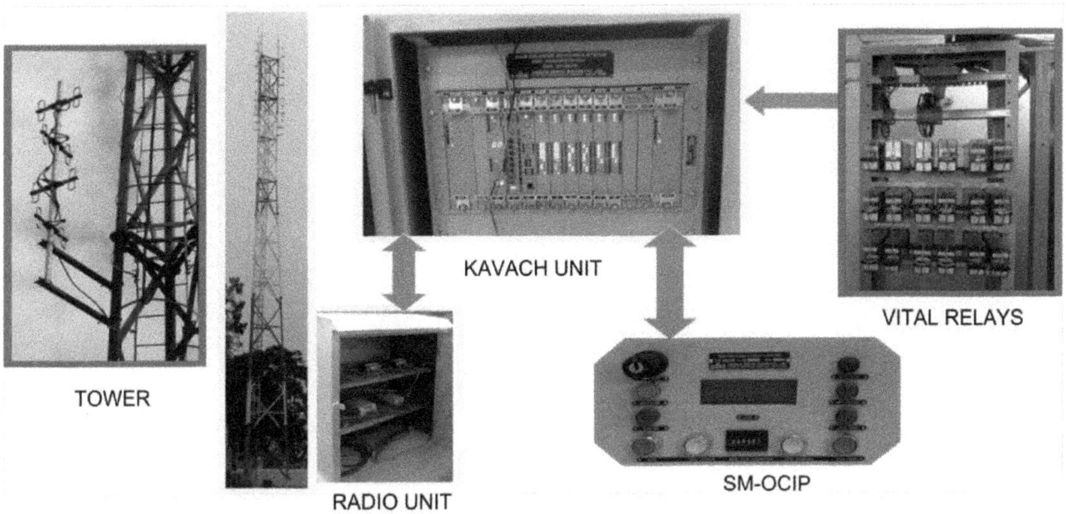

Fig. 1.11 Station KAVACH

1.16.2 On-Board Sub-Systems

The On-board subsystem shall be comprised

- Loco KAVACH Vital Computer
- RFID Reader
- Loco KAVACH Radio Unit with antennas and other communications
- Driver Machine Interface(DMI)
- Brake Interface Unit (BIU), where required

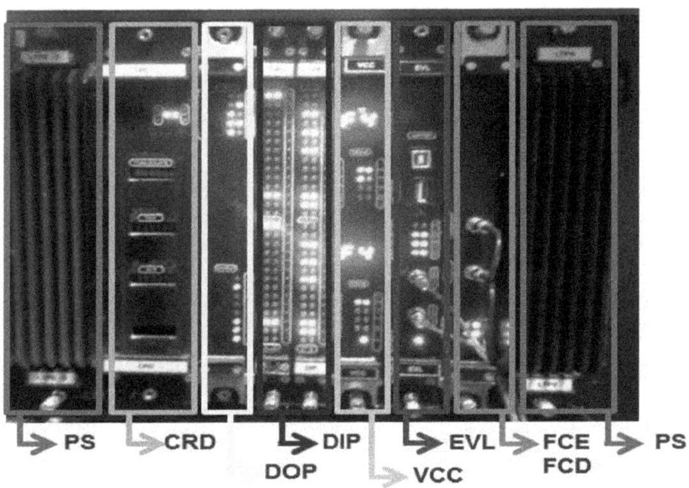

Fig. 1.12 A Loco Vital Computer

RFID Tag Reader

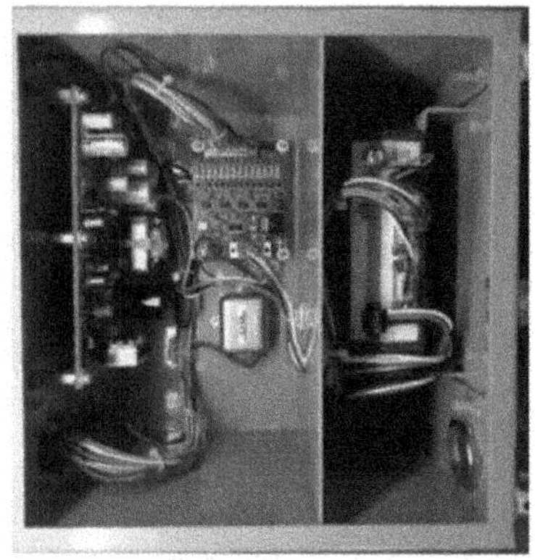

Loco Radio Modem

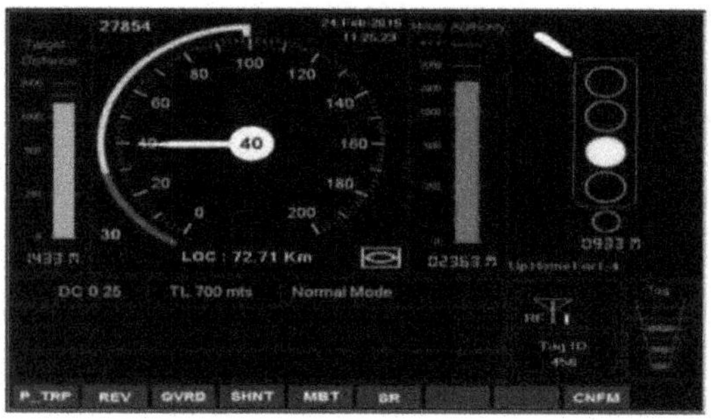

Driver Machine Interface (DMI)

Brake Interface Unit (BIU)

Fig. 1.12 B Loco KAVACH

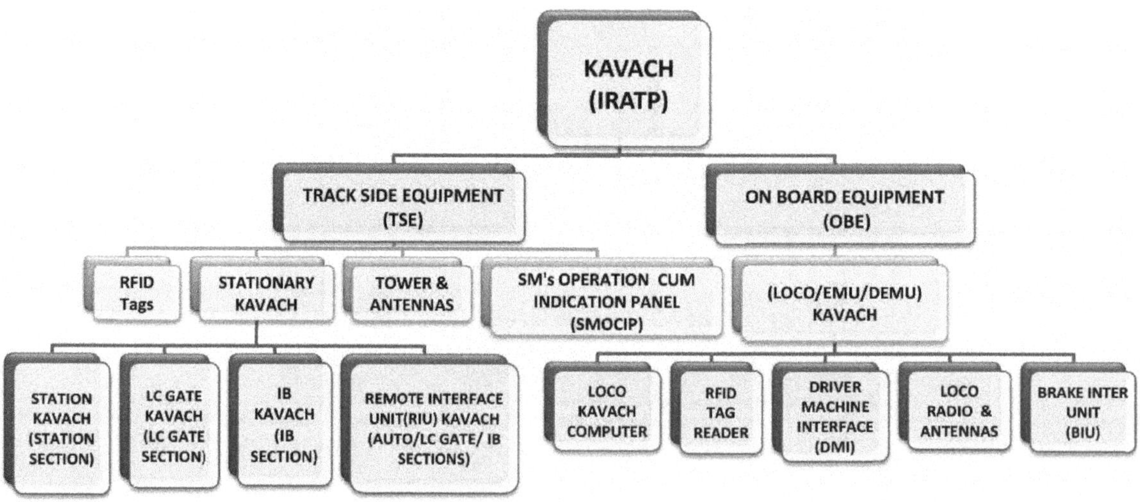

Fig. 1.13 Flow chart of KAVACH Sub-systems

1.17 Track Sides Sub-system

1.17.1 RFID Tag

RFID Tags provide location specific static information to Locomotive. Apart from acting as Location references these provide immediate information such as crossing the signal etc. to Loco Unit.

RFID tags are fitted on track in station section, point zones, near Signals & in block section for giving Trackside information to Loco KAVACH unit.

The RFID tags shall be fitted on the sleepers between the rails as per guidelines given for Indian Railways.

Specification of RFID tag :

- Suitable for reliable working at train speed upto 200 KMPH (minimum).
- Frequency of operation: 865-867 MHz.
- Can be programmable with minimum 128 bits (including CRC) of user data.
- Shall be able to work even when submerged in water up to rail level.
- Under field operating conditions RFID reader antenna shall be able to read,
- RFID tag from a vertical distance of 700 mm from bottom of RFID reader antenna to top of the rail level.

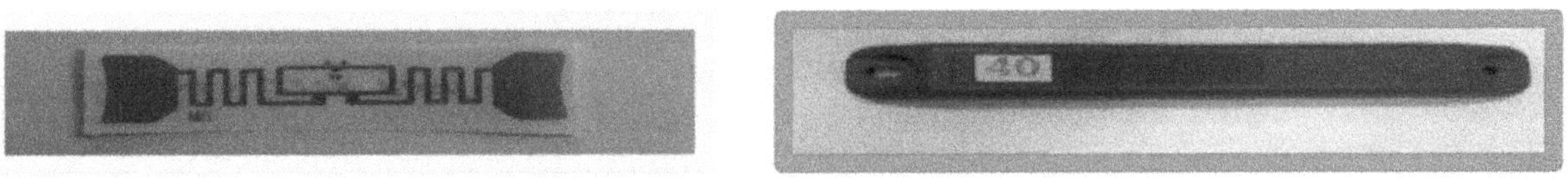

Fig. 1.14 RFID Tag fixed on sleeper

1.17.2 Stationary KAVACH Unit

Stationary KAVACH Unit shall be universally suitable for various types of signalling of Indian Railways with provision of colour light signalling. By default, it shall be suitable for interfacing with Panel Interlocking (PI), Route Relay Interlocking (RRI) and Electronic / Solid State Interlocking (SSI).

Normally Stationary KAVACH Unit shall be provided at Stations to cover all the trackside signals. It shall also be provided at Intermediate Block Locations (IBS) and midsection interlocked Level Crossing Gates where the radio signal coverage of station KAVACH tower is not adequate. This shall be interfaced with interlocking equipment to acquire real-time dynamic information related with signalling such as various signal aspects. It has database of static signalling related information such as location & details of RFID tags and Speed Restrictions. It gets real-time information regarding Locations, Speed etc of various trains in its jurisdiction through UHF Radio Communication.

On the basis of this information, it detects any emergency situation and can direct the command to Loco to take action to stop.

Stationary KAVACH Unit shall comprise of:

1. Station/ LC/ IB KAVACH Vital Computer
2. Stationary KAVACH Radio Unit
3. Remote Interface Unit (as per requirement)
4. Station Master Operation cum Indication Panel (SMOCIP)

1.17.2.1 Station/LC/IB KAVACH Vital Computer

The Vital Computer of Station/ LC/ IBS KAVACH Unit is a computer-based system that generates messages to be sent to the train on basis of information received from interlocking inputs and on basis of information exchanged with the Loco KAVACH units. Vital Computer architecture shall be minimum 2 out of 2. Station/ LC/ IBS Vital Computer shall have Real Time Clock synchronization facility with GNSS clock to synchronize with other KAVACH systems in hot standby manner. Station/LC/IBS Vital Computer shall have provision for the following:

- To interface with signalling inputs in fail-safe manner.
- Ethernet/ E1 port and two GSM interfaces for connectivity with Network Monitoring System (NMS) and Key Management System (KMS).
- To interface with OFC (E1 interface/ Dark Fibre) for connectivity with Remote Interface unit with minimum Four (six proposed).
- USB interface for downloading of log & other data for diagnostic purposes. To interface with Video Display Unit (VDU) to show real time display of Loco movements and signal aspects of the yard, (to be provided separately).

1.17.2.2 Stationary KAVACH Radio Unit

Radio communication network shall be used for the bi-directional exchange of messages between Loco KAVACH unit and StationaryKAVACH units.

Stationary KAVACH Radio Unit shall have two UHF full duplex Radio modems with separate cable and antennae in hot standby mode to communicate with Loco KAVACH unit.

1.17.2.3 Remote Interface Unit (RIU)

Remote Interface Unit (RIU) is a miniature version of Stationary KAVACH without radio communication unit, which captures (multiplexing) the relay information wired to it and exchanges the data with master Station KAVACH directly without any relay interfaces. Remote Interface Unit (RIU) shall be used where remote signalling functions are required to be fetched to a nearby Stationary KAVACH unit, for example from end cabin, distributed interlocking, a nearby LC Gate, Intermediate Block or Automatic Signalling section. RIU at such end cabins/ distributed interlockings or LC gate/ IB shall be installed if they are coming within the radio coverage of station tower. RIU is used to communicate remote signalling inputs to Stationary KAVACH over OFC media. In multiple RIUs scheme these are connected in a Ring network topology to increase the availability of the network. In a Ring network, each RIU is connected to two adjacent RIUs in primary and secondary. A single RIU shall be capable of communicating with two adjacent RIU units so that the operations are not affected in case of communication link failure on one side only. A single RIU shall be capable of handling at least 32 field inputs. RIU KAVACH unit shall consist of Vital Input modules with minimum Two-Out-Of-Two architecture.

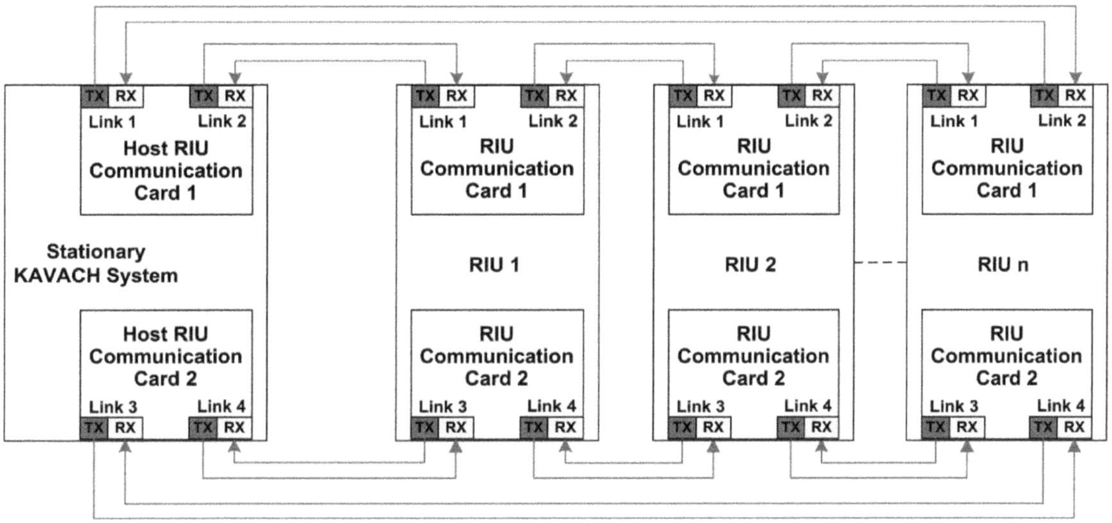

Fig. 1.15 Connectivity between Remote Interface Units (RIUs) and Stationary KAVACH

1.17.2.4 Station Master Operation Cum Indication Panel (SMOCIP)

Purpose of SMOCIP:

- Station Master can generate SOS to Loco KAVACH units in its jurisdiction.
- Cancellation of SOS generated by stationary KAVACH.
- Monitoring the health of Stationary KAVACH.
- SOS status display on LCD panel of SMOCIP.

Connectivity of SMOCIP:

- 12 Core signalling cable is used for SMOCIP in SM Room to Stationary KAVACH in Relay Room For button, counter & power supply
- 10 pair PIJF cable is used for SMOCIP in SM Room to Stationary KAVACH in Relay Room for communication between SMOCIP and Stationary KAVACH.

SMOCIP consists of following:

- LED Indications
- Switches
- LCD Panel
- Buzzer

- Digits counter which are updated on pressing of manual SOS switch Station Master's key.

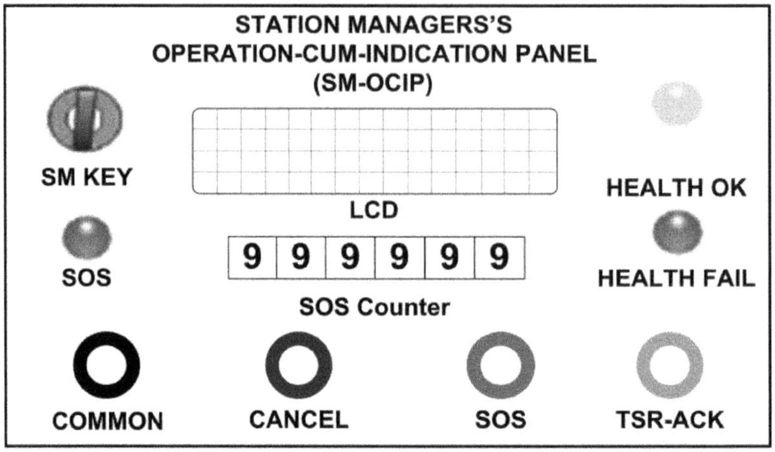

Fig. 1.16 Front Panel diagram of SMOCIP

It has the following LED Indications on the Console:

Sl. No.	LED Name	Colour	Description
1	HEALTH OK	GREEN	Indicates Stationary KAVACH Healthy
2	HEALTH FAIL	RED	Indicates Stationary KAVACH Un Healthy
3	SOS	RED	When SOS generated from Station

It has the following switches on the Console:

Sl. No.	LED Name	Colour	Description
1	COMMON	BLACK	Common switch to press along with SOS switch
2	CANCEL	BLUE	To cancel the SOS from station
3	SOS	RED	To generate SOS from Station

No functionality of switches is allowed without SM-KEY insertion into SM-OCIP.

Following indications/ buttons/ buzzers shall be given in the Station Master's OCIP:

(a) Station Masters Key

(b) LCD display (4 Line x 20 char)

(c) SoS indication

(d) Health indication

(e) Audio Buzzer

(f) Three Push Buttons (Common, Generation and Cancellation) to generate and cancel the SoS.

(g) Electromechanical non-resettable 6 digit counter for recording SoS operation.

1.17.3 Tower and Antenna

Self supported lattice towers of height 40meters are used to cater 4.5km range of radio coverage for Radio communication between Stationary KAVACH and Loco KAVACH.

The antennae for stationary communication system at station/ IBS/ midsection interlocked Gate unit shall be combination of vertically polarized omni and/ or directional antennae. The antenna cable & antenna shall be suitable to provide a minimum range of communication approximately 1.5 km on approach of first signal of the Stationary KAVACH unit (typically 4.5 kms in case of Double-Distant territory of Indian Railways).

1.18 On-board Sub-system (Loco KAVACH Unit)

The On-board subsystem shall be comprised of

(a) Loco KAVACH Vital Computer

(b) RFID readers consisting of two RFID Reader Antenna in hot standby

(c) Loco KAVACH Radio Unit consisting of two Radio Modems in hot standby with separate cables and antennae.

(d) Two Driver Machine Interface (DMI) for each locomotive or one DMI for each Driving motor coach of EMU/DMU/MEMU/DEMU etc.

(e) Brake Interface Unit (BIU).

1.18.1 Loco KAVACH Vital Computer

The Loco KAVACH vital computer is a system that supervises the movement of the train to which it belongs, on basis of information exchanged with Stationary KAVACH units and other Loco KAVACH units. Vital Computer architecture shall be minimum 2 out of 2. Loco KAVACH vital computer shall have Real Time Clock synchronization facility with GNSS clock to synchronize with other KAVACH systems in hot standby manner. Loco KAVACH vital computer shall have provision for the following:

- To interface with train interface unit & brake interface unit.
- Two Direction sensing type Speed Sensor interface for distance and speed measurement.
- To interface with RFID reader to read RFID tags fitted on the track.
- To interface with Driver Machine Interface (DMI) consisting of display arrangement & buttons/ switches for operation.
- Two GSM interfaces for connectivity with centralized Network monitoring System (NMS) and Key Management System (KMS). It shall be also to be operable with LTE where LTE is proved.
- USB interface for downloading of log &other data for diagnostic purposes.

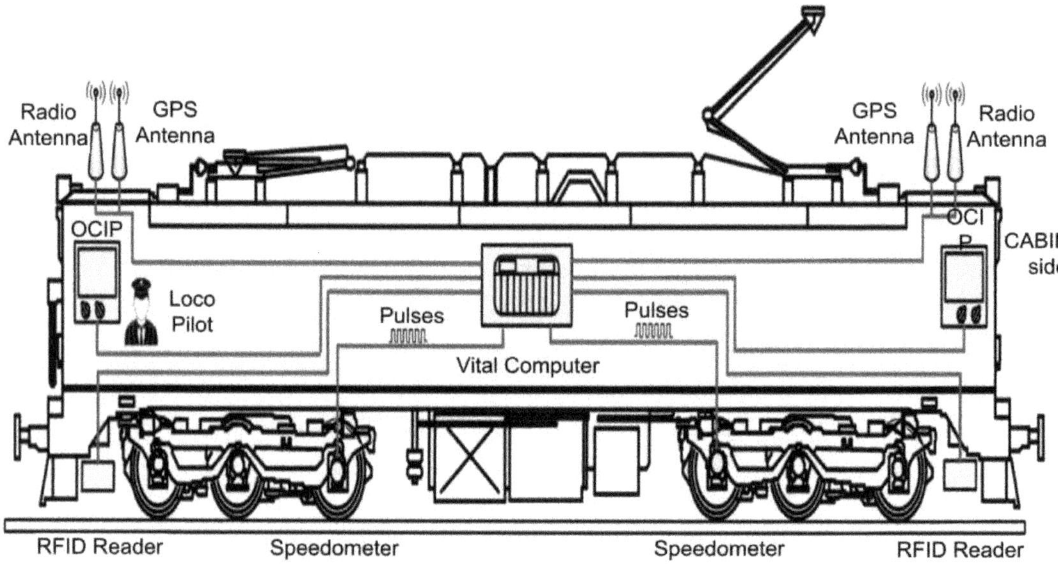

Fig. 1.17 A Electric Loco KAVACH

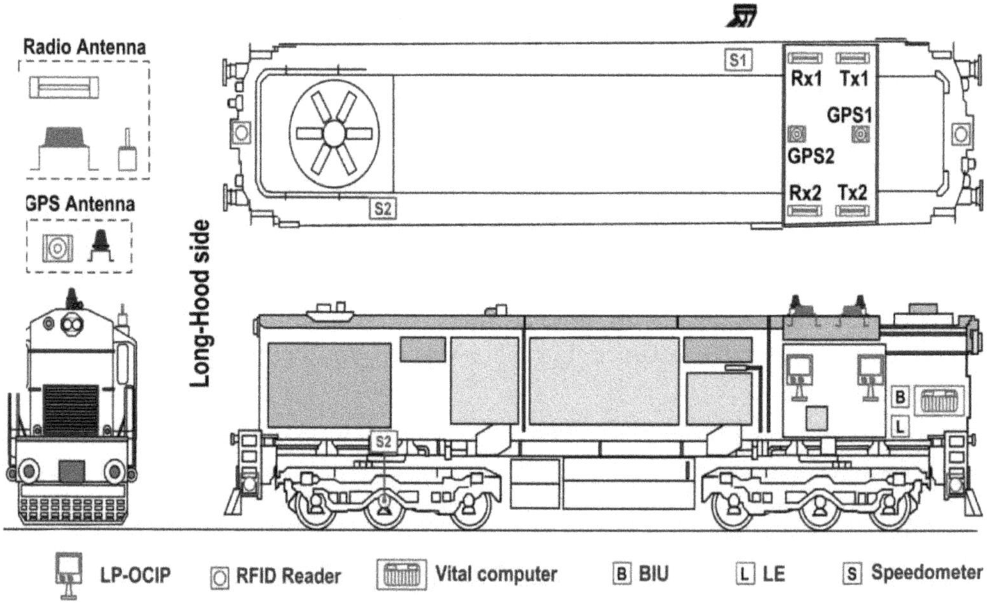

Fig. 1.17 B Diesel Loco KAVACH

1.18.2 RFID Reader

Each Loco KAVACH unit shall have two RFID readers for getting the information from RFID tags fitted on the trackside in hot standby manner.

1.18.3 Loco KAVACH Radio Unit

Loco KAVACH radio unit specifications shall be similar to stationary KAVACH radio unit. Loco KAVACH radio unit shall have two uhf full duplex radio modems with separate cable and antennae in hot standby mode to communicate with stationary KAVACH unit.

1.18.4 Loco Pilot Machine Interface (DMI)

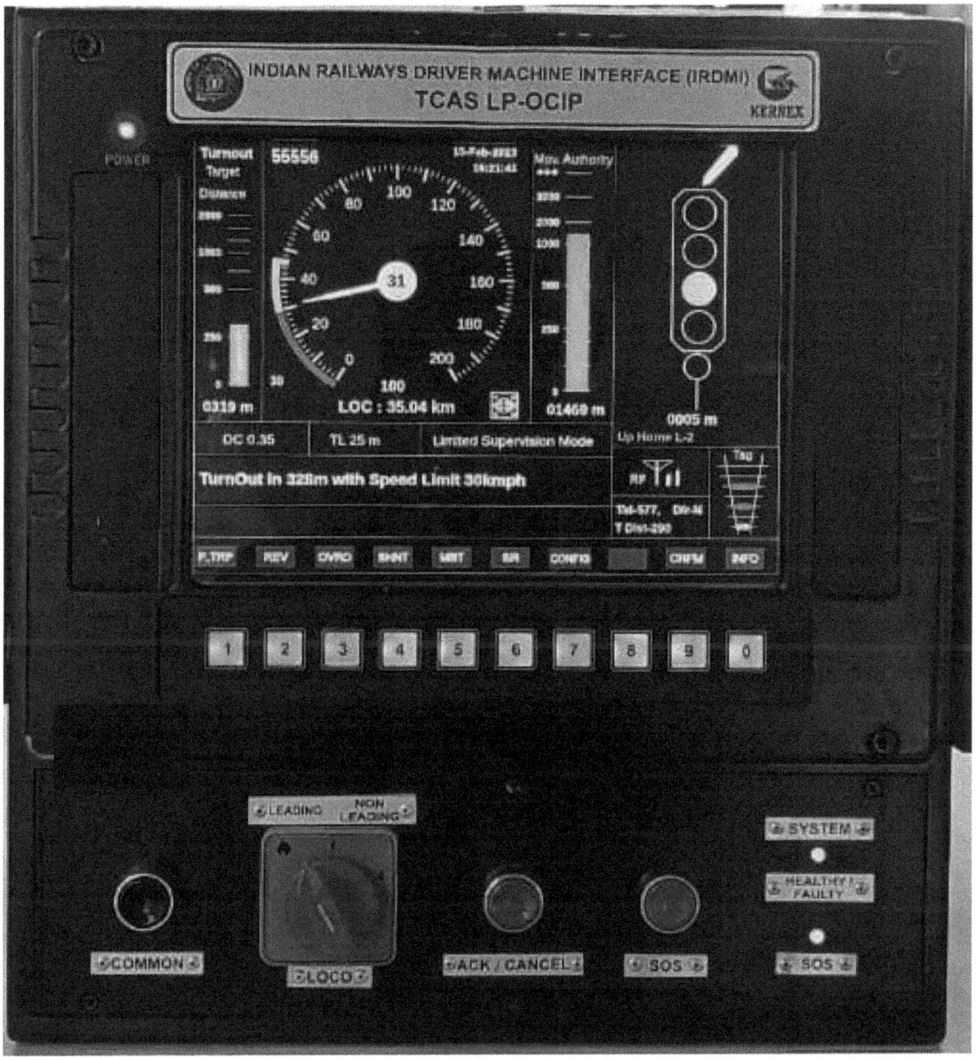

Fig. 1.18 A DMI

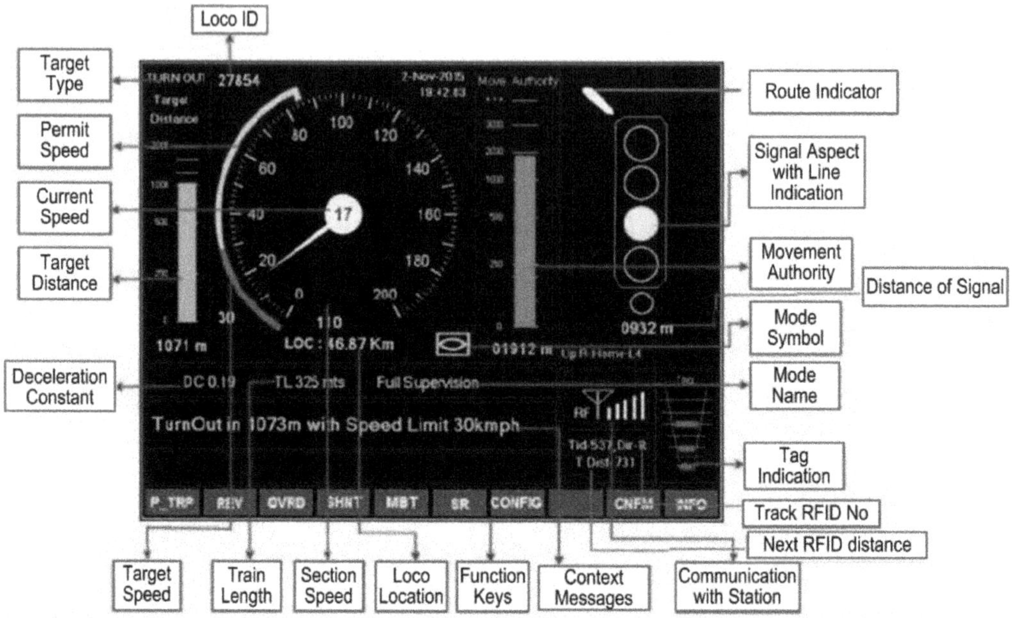

Fig. 1.18 B

Loco Pilot Machine Interface (DMI) is a general, operational, system, technical, functional, and performance requirements for Loco KAVACH Driver Machine Interface so that there is clear and consistent understanding between the Loco Pilot and the KAVACH system.

Most of the information displayed is in form of Analog Displays in form of Circular Gauges (Arcs) and bars keeping ergonomics and convenience of Loco Pilots. Digital values have also been additionally displayed. Audio prompts and warnings also are implemented.

Two LP-OCIPs (Operation cum Indication Panel) provided for Loco pilot and Asst. pilot interaction with KAVACH.

Loco Pilot's Operation-cum-indication panel (LP-OCIP/DMI) shall consist of suitable arrangements and buttons/ switches for display/ operation of following functions:

(a) Communication with Loco KAVACH.
(b) Train type selection by the loco pilot.
(c) SOS operation by the loco pilot.
(d) Signal aspect display.
(e) Train length display
(f) Train type display
(g) Display of all modes of loco operation
(h) Current speed

(i) Over speed
(j) Permitted speed
(k) Target speed (For entering into loop line)
(l) Movement Authority (MA)
(m) The function in DMI for displaying the context messages for Loco Pilot's attention

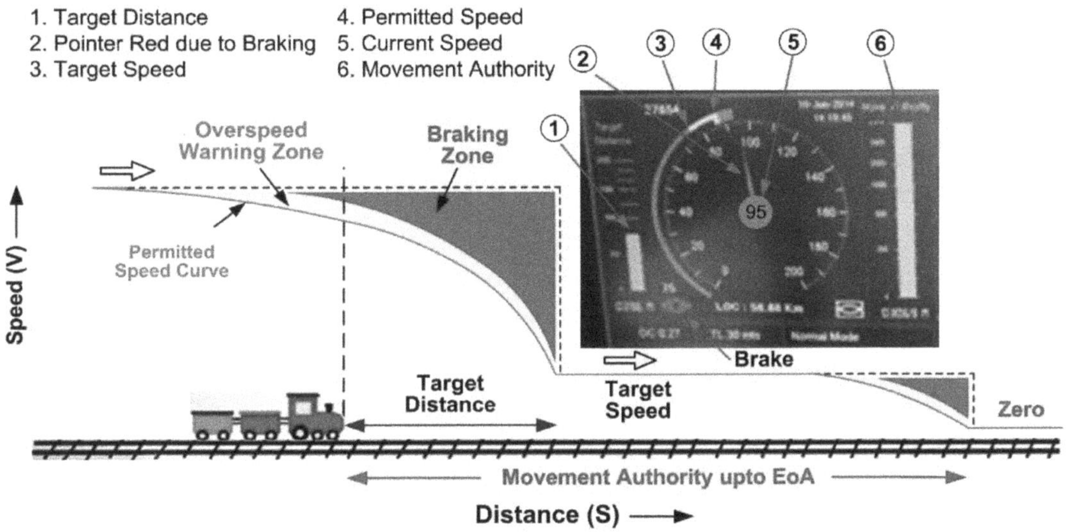

Fig. 1.19 A DMI with Movement Authority

Fig. 1.19 B DMI Showing Over Speed

Braking Systems in Railway Vehicles

I. Introduction

The brakes are used on the coaches of railway trains to enable deceleration, control acceleration (downhill) or to keep them standing when parked. While the basic principle is similar from road vehicle, the usage and operational features are more complex because of the need to control multiple linked carriages and to be effective on vehicles left without a prime mover. In the control of any braking system the important factors that govern braking action in any vehicle are pressure, surface area in contact, amount of heat generation and braking material used. Keeping in view the safety of human life and physical resources the basic requirements of brake are:

The brake must be strong enough to stop the vehicle during an emergency with in shortest possible distance.

There should be no skidding during brake application and driver must have proper control over the vehicle during emergency.

Effectiveness of brakes should remain constant even on prolonged application or during descending on a down gradient

Brake must keep the vehicle in a stationary position even when the driver is not present.

The brake used in railway vehicles can be classified according to the method of their activation into following categories.

(a) Electrodynamic Brake

(b) Mechanical Brake

(c) Electromagnetic Brake

(d) Pneumatic Brake

(e) Compressedair Brake

(f) Vacuum Brake

(a) Electrodynamic Braking System: Braking system used is electric trains is electrodynamic braking that converts the motor into a braking generator dissipating the kinetic energy in the form of heat. Regenerative braking uses the generated electricity instead of dissipating it as heat, and is becoming more common due to its ability to save energy. Principle of the electrodynamic traction, dynamic braking and regenerative braking systems is shown in Figures below.

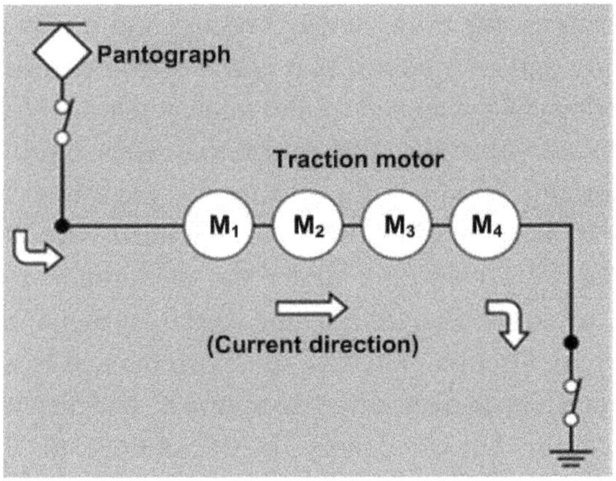

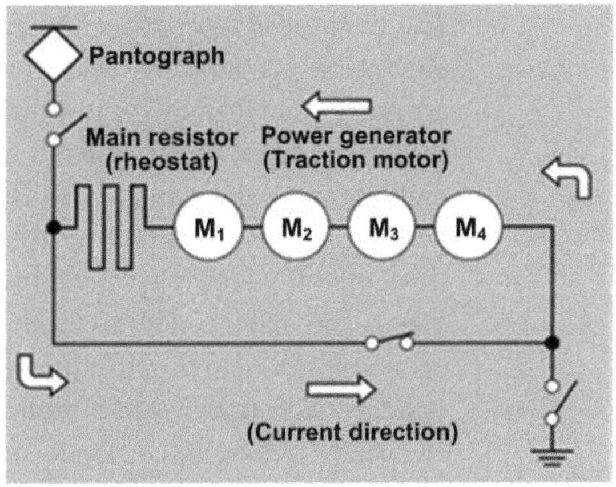

Fig. 1.20 Principle of electrodynamic traction **Fig. 1.21** Principle of dynamic braking

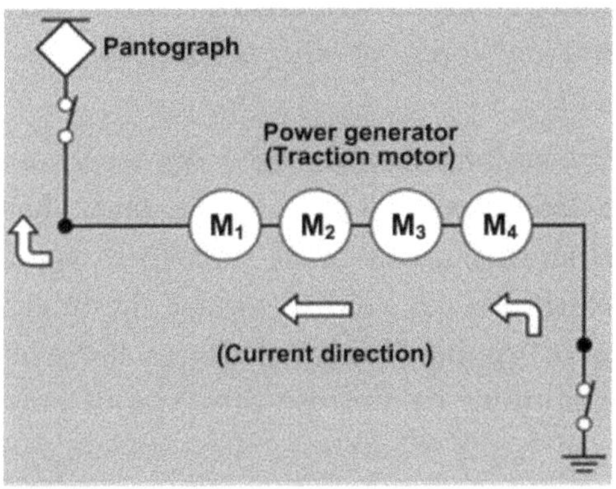

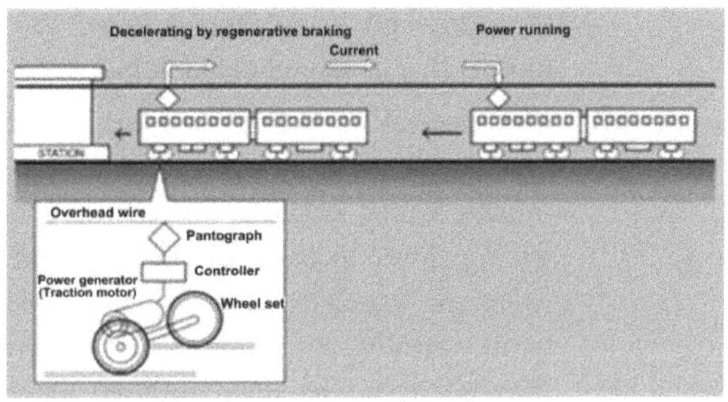

Fig. 1.22 Principle of regenerative braking **Fig. 1.23** Principle of recycled regenerated electric power

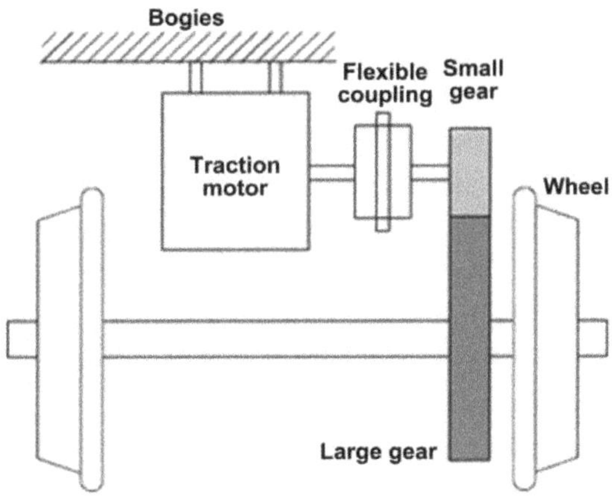

Fig. 1.24 Transmission of braking force from traction motors to wheels

The traction motor drives and accelerates the train, during braking and it acts as an electric generator instead, forming part of a circuit that consists of a rheostat, armatures and a field system. Electricity is consumed by the main resistor, which converts kinetic energy of the train into heat and acts as a brake. Regenerative braking uses the same type of circuit; however the electricity generated by braking is not consumed by rheostat. It is transmitted to the overhead wire.The flow of this electricity is controlled by a controller under the pantograph that opens and closes within fraction of time. Electrodynamic brake systems are economical to use because they do have friction elements, as in mechanical brake systems. The regenerative braking system is even more economical because the electricity regenerated from the trains kinetic energy is transmitted to the overheadwire, and becomes available to power other rolling stock.

However electrodynamic brake systems occasionally malfunction because they have complex circuits. Therefore they cannot be used as emergency brakes. In an electrodynamic braking system, the braking force of the traction motor is transmitted to the wheels via gears.

(b) **Mechanical Braking System:** The basic braking devices used by mechanical braking systems are: wheel tread brakes, axle-mounted disc brakes, and wheel-mounted disc brakes. These brake mechanisms use a brake shoe that applies friction force to the disc. The applied pressure is adjusted to control the braking force. In wheel-tread brake, the brake shoe applies friction force to the wheel tread, creating a sliding effect. High-speed trains cannot use this type of brake, because doing so may damage the wheel tread. Therefore, they use axle- or wheel-mounted disc brakes. Axle-mounted disc brakes require sufficient space to accommodate therefore used in trailer bogies. Wheel-mounted disc brakes are used on motor bogies because it requires accommodating the traction motor only and having insufficient space for an axle-mounted brake. In both systems, compressed air or oil is applied to a brake cylinder that pushes the brake lining against the disc. Brake discs are dead weight that is useful only during braking, therefore operators can install lighter discs. Carbon/carbon-composite multi-discs and aluminium composite discs offer lighter weights and are widely used. The carbon/carbon-composite multi-disc has alternate sections of carbon-fiber rotors and stators. During braking, they rub against each other to create a frictional force that slows down the wheel or axle. The disc is lighter in weight than conventional materials and has good heat-resistant properties. Aluminium-composite brake discs may be made much lighter than today's forged steel and cast-iron brake discs. Moreover their structure is common for both axle or wheel-mounted discs, achieving a much lighter disc without design.

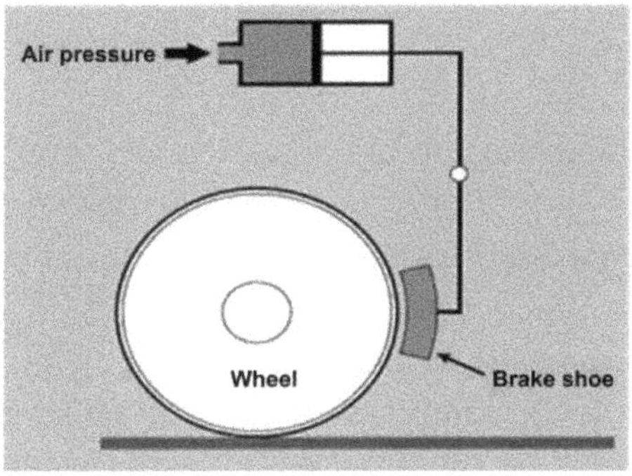

Fig. 1.25 Principle of wheel tread brakes

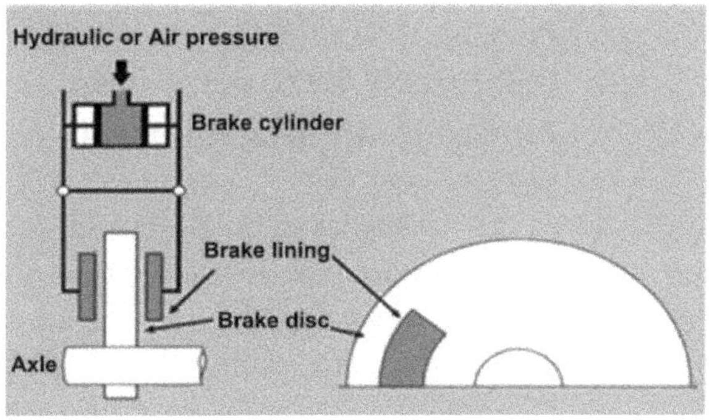

Fig. 1.26 Principle of axle-mounted disc brakes

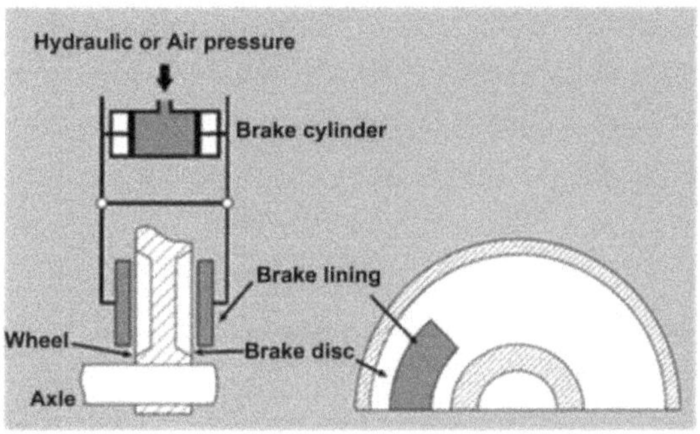

Fig. 1.27 Principle of wheel-mounted disc brakes

Fig. 1.28 Carbon/Carbon-composite multi-disc system

(c) **Electromagnetic Braking System:** Conventional train braking systems depend heavily on adhesion between the wheel tread and the rail. In the case high-speed trains, adhesion decreases as speeds increase, making it necessary for the train to reduce braking force to avoid wheel sliding. This result is longer braking distances. To overcome this problem, a electromagnetic brake system that does not depend on adhesion was developed. It produce a braking force by using magnetic repulsion obtained from eddy currents generated on the top surface of the rails. Earlier it was not used because of assumption that the eddy currents would heat small sections of the rail to such a degree that the rail would bend sideways. This is solved by development of a electromagnetic brake that uses eddy currents and frictional force. The electromagnetic brake on bogie is connected to batteries that create alternating north and south poles forming magnetic fields between the poles. The magnetic fields generate eddy currents in the top surface of the rails, creating a force acting in an opposite direction to the movement of the train, in other words, a braking force.

(d) Pneumatic Brake:

(i) Compressed Air Brake Systems

Automatic Air Brake System: An automatic air brake system is shown in Fig. Air compressors mounted every two to four coaches supply compressed air to the air brakes. The air, which is compressed to nearly 8 kg/sq.cm, is piped below coach floors to main air reservoirs. The air pressure is lowered to 5 kg/sq cm with pressure regulator and air is fed via the brake valve, brake pipes, and control valves to auxiliary air reservoirs. If the compressed air in the brake pipes and auxiliary air reservoirs of each coach is at 5 kg/sq cm, brakes are not activated. The activated brake valve cuts the flow of air from the pressure regulator and air pressure in the brake pipes falls. The fall in air pressure is detected by the control valves on each coach. The control valves then regulate the flow of compressed air from auxiliary air reservoirs to brake cylinders. The brake cylinders activate the basic braking mechanisms to slow down and stop the coach. The control valves regulate the flow of air from the auxiliary air reservoirs to the brake cylinders at a pressure that is proportional to pressure drop in the brake pipes.

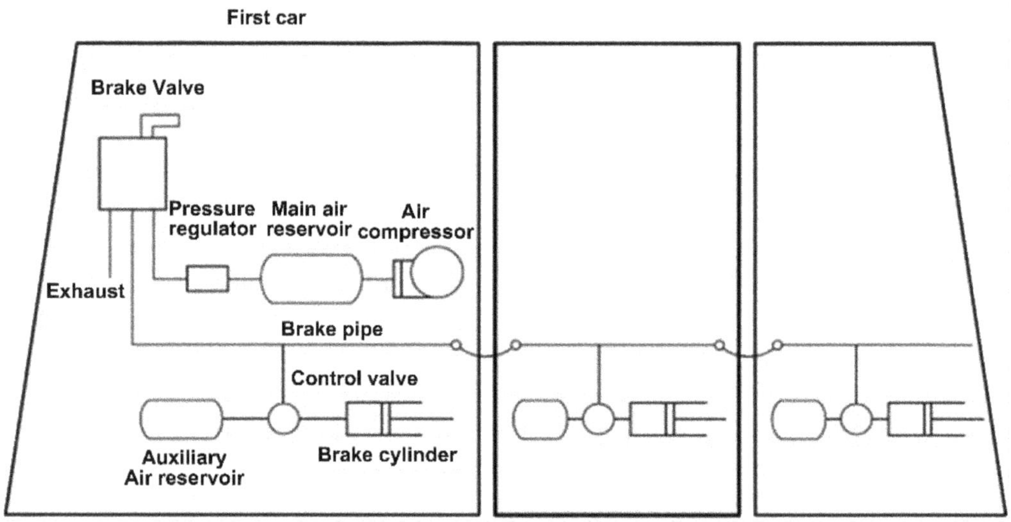

Fig. 1.29 Principle of automatic air brake system

(b) Straight Air Brake System: A straight air brake system is shown in Fig. 1.30. The straight air brake system does not have a control valve or auxiliary air reservoir in each coach as in automatic air brake system. Activation of brake valve forces compressed air from straight air pipe to brake cylinders, activating the basic braking mechanism. As the straight air pipes do not contain compressed air during normal running conditions, the brakes would fail if coaches became uncoupled. In order to avoid this, the straight air brake system may be used in conjunction with the automatic air brake system. It can also be avoided by using another pipe, called a main air reservoir pipe, from the first to the last coach. The air pressure in main air reservoir pipe acts like the compressed air in the brake pipes of the automatic air brake system. If compressed air in this main air reservoir pipe falls, or if it leaks from air pipes or from air hoses between coaches, etc., pressure drop is detected and brakes are applied automatically.

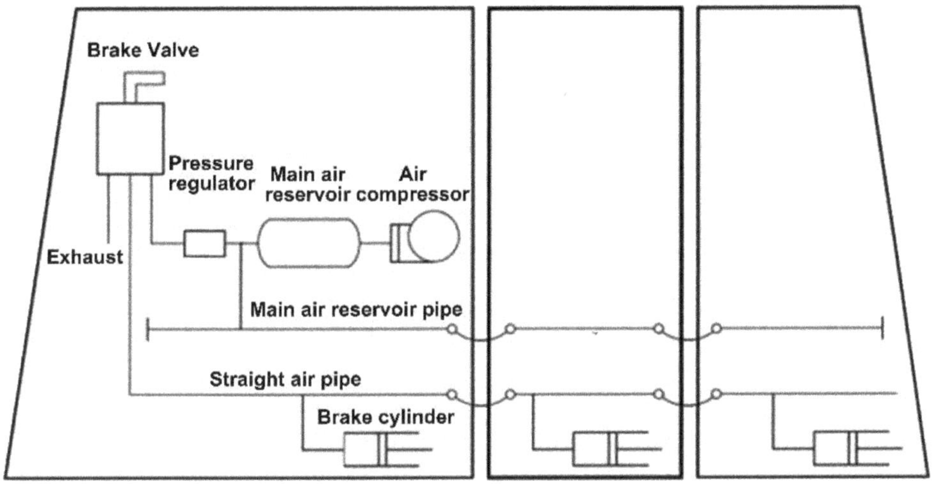

Fig. 1.30 Principle of straight air brake system

Air brake system may also be classified as follows:

Direct release air brake system

Graduated release air brake system

Direct release air brake system is most suitable for leveled track or constant gradient route. Due to this reason it is not suitable for Indian Railways. Graduated release air brake system is most suitable for Indian Railways. In graduated release air brake system the brake pressure is applied and released such that the magnitude of braking force is proportional to reduction in brake pipe pressure.

Graduated release air brake system can also be divided into two categories.

Single pipe graduated release air brake system

Twin pipe graduated release air brake system

(c) **Single Pipe Graduated Release Air Brake System:** Single Pipe Graduated Release Air Brake System is shown in Fig. 1.31. The operation is same as that of the twin pipe system except that the auxiliary reservoir is charged through the D.V. instead of feed pipe, since there is no feed pipe in single pipe system. As compared to single pipe graduated release air brake system, twin pipe graduated release air brake system is more suitable for passenger coaches.

(d) **Twin Pipe Graduated Release Air Brake System:** In twin pipe graduated release air brake system (Fig. 1.32), The Brake pipe is charged to 5 kg/cm^2 by the driver's brake valve. The auxiliary reservoir is charged by the feed pipe at 6 kg/cm^2 through check valve and choke. The brake cylinder is connected to the atmosphere through a hole in the D.V. when brakes are under fully released condition. To apply brakes, the driver moves automatic brake valve handle either insteps for a graduated application or in one stroke to the extreme position for emergency application. By this movement the brake pipe pressure is reduced and the pressure differenced is sensed by the D.V. against the reference pressure locked in the control reservoir. Air from the auxiliary reservoir enters the brake cylinder and the brakes are applied. At the time of release the air in the brake cylinder is vented progressively depending upon the increase in the brake pipe pressure. When the brake pipe pressure reaches 4.8kg/cm^2 the brake cylinder is completely exhausted and brakes are fully released.

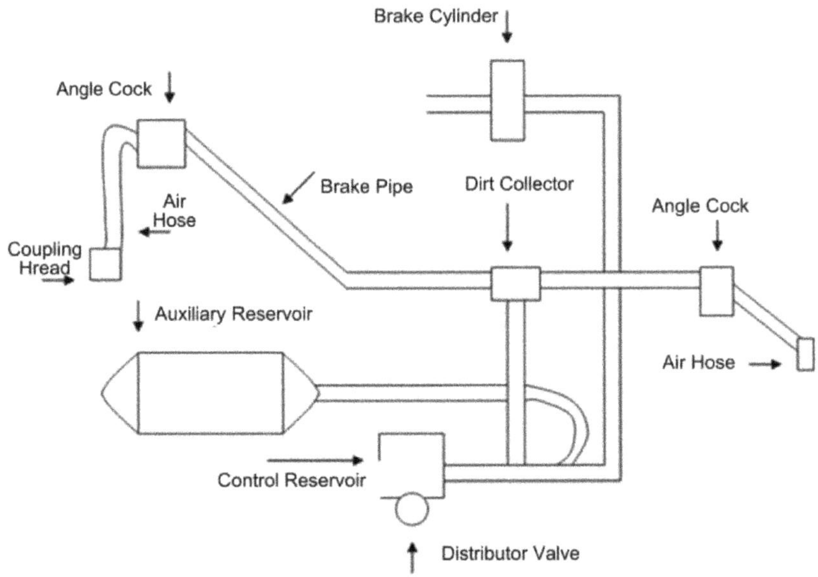

Fig. 1.31 Single pipe graduated release air brake system

II. Calculation of Train Stopping Distance

For trains to safely travel on a railway, trains must be provided with sufficient distance in which to stop. Allowing too long a distance reduces the capacity of the line and has an impact on rail infrastructure investment. Too short a distance and collisions would occur, because the train would not be able to stop with in the available distance and would therefore occupy a section of track that could be allocated to another train.

Consequently it is important that distance be adequate. Train braking distance is function of following factors

Train speed when the brakes are applied.

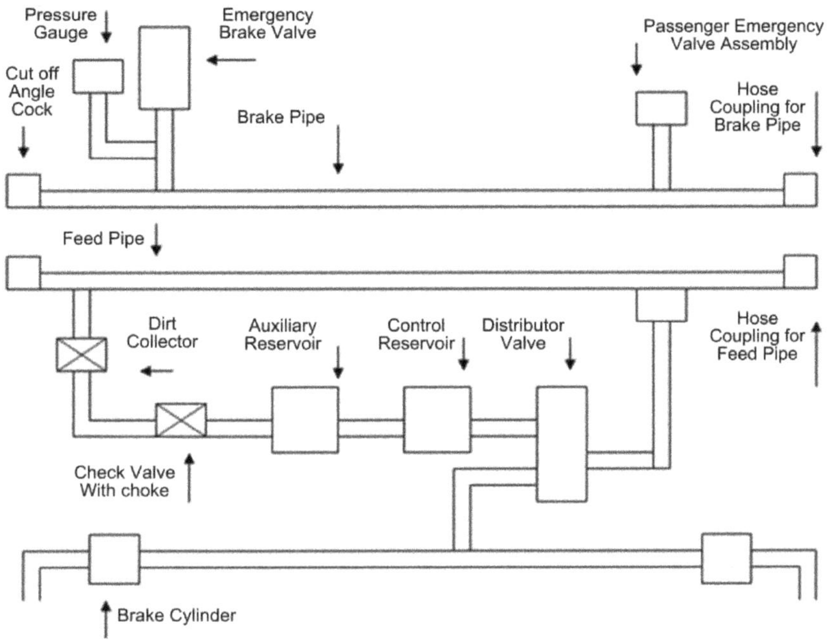

Fig. 1.32 Twin pipe graduated release air brake system

The available friction at wheel-rail surface which influences the retardation rate for complete brake application.

Time from when the brakes are applied by the train driver to when they are actually become effective i.e. brake delay time.

The magnitude of wear of brake pads and the pressure available in brake cylinders.

Track gradient when brakes are applied and mass distribution of track.

In order to stop the train it requires the work. The required work is the sum of change in the train's kinetic energy and the change in its potential energy due to change in the height due to the gradient of the track.

Mathematically it may be expressed as:

$$mas + \frac{mV^2}{2} + mg(h_2 - h_1) = 0 \qquad .....(1)$$

Where,

M = mass of train,

V = Speed at which the retardation begins

S = Stopping distance,

h_1 = Height at which the retardation begins,

h_2 = Height at which the train stops ($h_2 - h_1$)
a = Retardation provided by braking system,

The above equation suggests that mass has no direct effect on the train stopping distance. However mass distribution has influence on train stopping as train's centre of gravity varies with the mass distribution. In case off reight wagons where the mass varies from no load to full load the rear two levels of brake for ceused empty and loaded.This influences the design of the brake system. For calculating the braking distance calculations the lowest deceleration rate is used to calculate the deceleration rate for the complete train.

Eq. 1 may be written as

$$S = \frac{(-V^2)}{2(a - g\tan a)}, \text{ for } a < 0 \quad .....(2)$$

α = Angle of slope, for small values of α, $\sin\alpha \approx \tan\alpha$

Assuming constant gradient track and considering brake delay time the stopping distance can be calculated using following expression.

$$S = \frac{(-V + bt_d)^2}{2(a+b)} - Vt_d - \frac{bt_d^2}{2} \quad .....(3)$$

b = Retardation provided by gravity; t_d = Brakedelaytime

1.19 Brake Interface Unit (BIU)

BIU shall apply normal, service & emergency brakes of locomotives respectively based on the type of brake command received from Loco KAVACH unit. In addition to these brakes, it shall also apply Loco brake, if Loco Brake command is generated by Loco KAVACH.

S.No.	Name of Brake	Symbol
1	Normal Brake	
2	Full Service Brake	
3	Emergency Brake	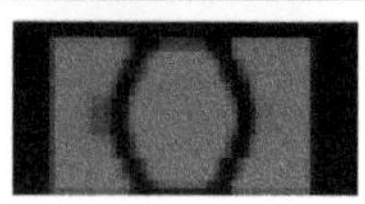

Fig. 1.20 NB, FSB & EB

1.19.1 BIU consists of Two Modules

1. **Electronic Module:** It consists of a Control Card having built-in air brake control logic. Control Card interfaces with Analog Input Module, Digital Input Module & Digital Output Module to monitor and control air brake application/ release.
2. **Pneumatic Panel:** It consists of different valves, pressure transducers and manual cocks-Interfaced with IRAB system & Electronic Module for brake control purpose.

1.19.2 BIU of Locomotives

At present IR is using different types of Electric and Diesel Locomotives. In a broad way, three types of brake systems Viz. IRAB, CCB and E-70 brake systems are being used in these locomotives. Since the design of all these brake systems is different, the KAVACH-Specific BIUs are also different for each brake system. The classification of various BIUs specific to brake system is given below.

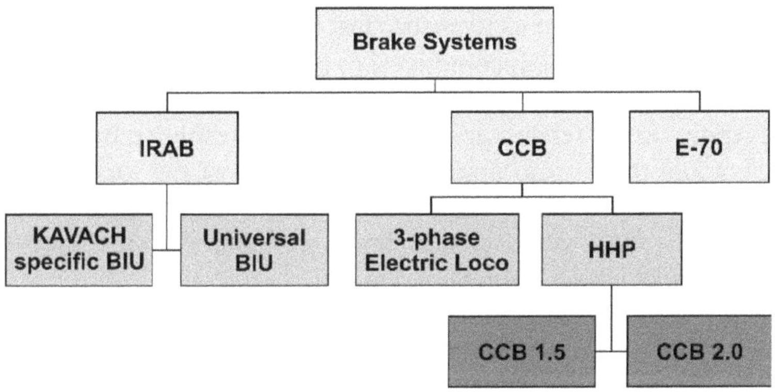

1.19.3 BIU for IRAB Brake System

1. To understand the working of Brake Interface Unit, the working of BIU for IRAB brake system is explained in this section.
2. In the locomotives to enable the application of brakes in the locomotive and formation, locomotive generates and maintains a pressure of around 10kgs/cm^2 in MR (Main reservoir) and this pressure is reduced by the pneumatic system and fed throughout the brake pipe (BP) and feed pipe (FP). The pressure in BP is maintained at 5.0 kg/cm^2 and FP is maintained at 6.0kg/cm^2. The purpose of FP is to enable quick release of brakes and hence there is no need to interface with KAVACH BIU. To apply the brake while working a train (Locomotive + Formation), Loco Pilot destroys pressure in brake pipe to the desired level and based on the reduction of pressure in Brake Pipe, brakes are applied in the train. The greater the reduction of Brake Pipe pressure, the stronger the application of brake i.e more Brake cylinder pressure . To ensure the complete release of brakes in the train, Loco Pilot again needs to maintain the B.P pressure at 5.0kg/cm^2 by keeping the handle in release position.

3. In IRAB brake system, there are two types of brakes provided for usage i.e. through A9 (train brake) and through SA9 (locomotive brake).
4. The A9 valve (Train Brake) is having four positions viz. Run/release, Initial, Full Service and Emergency. To create BP to 5.0kg/cm^2 loco pilot keeps the A9 handle in run position. This will ensure the complete release of brakes in the Train.
5. When Loco Pilot, places the A9 handle (of A9 valve) in Initial, Full service or Emergency position, BP reduces through the IRAB brake system to 4.5kg/cm^2 (0.5 kg/cm^2 reduction), 3.5kg/cm^2 (1.5 kg/cm^2 reduction) or 0kg/cm^2 (5.0 kg/cm^2 reduction) as the case may be to ensure application of brake in the formation. The mechanism of application of brake in the formation as a result of reduction in Brake pipe pressure is beyond the scope of present discussion. (This mechanism is part of IRAB brake system).
6. In addition to the above, during emergency position, the brake pipe pressure is vented to atmosphere by direct vent in the A9 vale to ensure quick reduction of brake pipe pressure so as to ensure fast application of brakes.
7. While working single locomotive i.e. as LE (light engine), the brake application in LE takes place through SA9 valve. This valve has two positions i.e. Release and Apply. In the release position, SA9 valve results in brake cylinder pressure of 0kg/cm^{2}and there is no application of brakes in the locomotive. When the SA9 handle is operated from Release to Apply (depending on the position of SA9 handle), the SA9 valve creates brake cylinder pressure upto 3.5kg/cm^2 which results in application of brake in the LE.

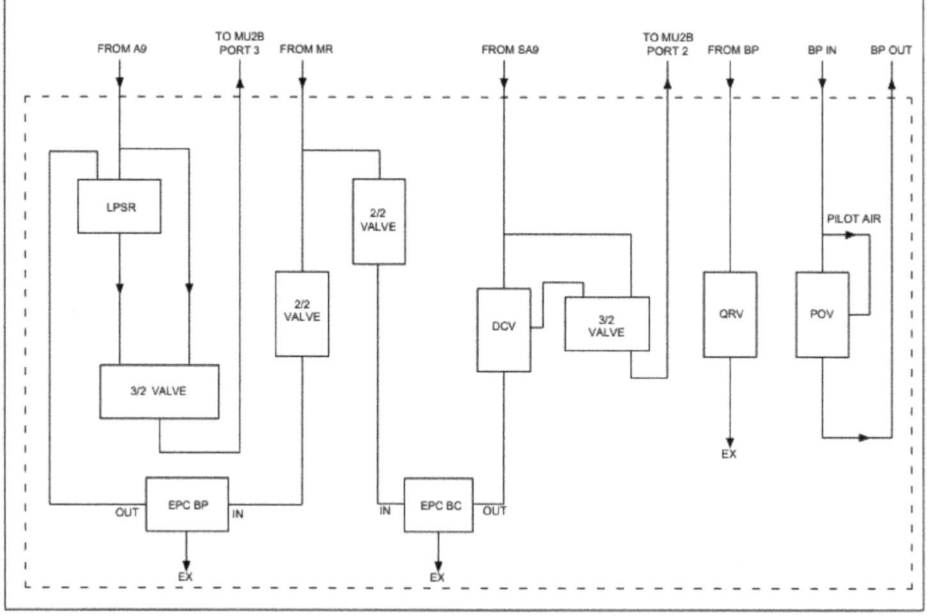

BIU -IRAB PNEUMATIC CIRCUIT

Working of IRAB Brake Interface Unit

Application of Train Brake through KAVACH BIU

From the above basics, if the KAVACH need to apply brake while working as Train (Locomotive + Formation), it has to perform the similar job of operating A9 handle to ensure application of Normal Brake/Full Service Brake(FSB)/Emergency Brake(EB) as the case may be.

The block diagram of BIU of IRAB brake system is shown above. For application of brake through A9 circuit, BIU of KAVACH is integrated into the A9 circuit of locomotive through LPSR(Low pressure Sensor), EPC-BP(Electric Pneumatic Controller for BP), 3/2 Valve and 2/2 valve. The EPC-BP, 3/2 Valve and 2/2 valve are electrically operated valves through KAVACH electronic module. Whenever the KAVACH system is working, it first operates 2/2 valve which is energized to open. The input for 2/2 valve is MR (Main Reservoir) pressure of locomotive which is maintained at 10kg/cm^2 and output is connected to EPC-BP. When the electrical supply cut off, then 2/2 valve isolates MR pressure from feeding EPC-BP. EPC-BP is a electrically controlled valve which will reduce the input pressure of MR to a desired level based on the electrical current given to controller. It takes current from 4-20mA and depending upon the extent of current out pressure will be created i.e. higher the input current higher will be the pressure output. When the current signal is withdrawn, it will exhaust the pressure in output pipe line through exhaust port (shown as Ex). The 3/2 valve is also an electrically operated valve which can route the two input i.e. one direct input from A9 (LP command) or output from LPSR to the output port. When electrical supply is given to 3/2 valve, the LPSR output pressure is routed to output or otherwise input from A9 is routed to output. During isolation of KAVACH, the 3/2 valve is de-energized and the LP commands are directly routed through its output and KAVACH brake commands through A9 are isolated.

During working of KAVACH, the 3/2 valve is energized such that the output of LPSR is given as input to the IRAB brake system. LPSR is a pneumatic valve which allows the lowest of the input pressures (Two different pressures) to its output port. 2/2 valve is also energized to provide input pressure to EPC-BP. When KAVACH is not intervening with brake system, the current signal to EPC is kept such that it always creates 5.5kg/cm^2. This will ensure that output of LSPR always corresponds to A9 valve output of LP (the maximum value of A9 is 5.0 kg/cm^2) since it is lowest as compared to output of EPC-BP and enables application of brakes by LP.

When KAVACH initiates braking, it provides appropriate current signal to EPC-BP such that it creates 4.5kg/cm^2, 3.5kg/cm^2 or 0kg/cm^2 as the case may be for Normal/FSB/Emergency brake application through KAVACH. In addition to the above, LPSR functionality also allows LP to apply any more restrictive brake than KAVACH as it always allows lowest of the pressures to its output.

In addition to the above, KAVACH while applying Emergency Brake, de-energizes QRV (Quick Release Valve) to vent out BP to atmosphere similar to the provision in

A9 handle. The normal condition of QRV is always energized except during emergency brake application through KAVACH.

Application of Loco Brake through KAVACH BIU

When KAVACH has to apply brake while working as LE (Light Engine), it has to perform the similar job of operating SA9 handle to ensure generation of brake cylinder pressure.

The block diagram of BIU of IRAB brake system is shown above. For application of brake through SA9 circuit, BIU of KAVACH is integrated into the SA9 circuit of locomotive through DCV(Double Check Valve), EPC-BC (Electric Pneumatic Controller for BC), 3/2 Valve and 2/2 valve. The EPC-BC, 3/2 Valve and 2/2 valve are electrically operated valves through KAVACH electronic module. Whenever the KAVACH system is working, it first operates 2/2 valve which is energized to open. The input for 2/2 valve is MR (Main Reservoir) pressure of locomotive which is maintained at 10kg/cm^2 and output is connected to EPC-BC. When the electrical supply cut off, then 2/2 valve isolates MR pressure from feeding EPC-BC. EPC-BC is an electrically controlled valve which will reduce the input pressure of MR to a desired level based on the electrical current given to controller. It takes current from 4-20mA and depending upon the extent of current, out pressure will be created i.e. higher the input current higher will be the pressure output. When the current signal is withdrawn, it will exhaust the pressure in output pipe line through exhaust port (shown as Ex). The 3/2 valve is also an electrically operated valve which can route the two input i.e. one direct input from SA9 (LP command) or output from DCV (Double Check Valve) to the output port. When it is electrical supply is given to 3/2 valve, the DCV output pressure is routed to output or otherwise input from SA9 is routed to output. During isolation of KAVACH, the 3/2 valve is de-energized and the LP commands are directly routed through its output and KAVACH brake command through SA9 circuit are isolated.

When KAVACH initiates, FSB/EB through SA9 circuit it creates EPC output pressure of 2.5kg/cm^2 or 3.5kg/cm^2 as the case may be routed through DCV (Double check valve) for application of brake through SA9. The double check valve output is always maximum of the two input pressures connected to it. In case LP applies higher brake through SA9 than KAVACH, then such higher pressure will be routed into SA9 circuit as output of Double Check Valve is maximum of two input pressures connected to it.

In this way, KAVACH Interfaces its brake commands through BIU.

1.20 Working of BIU

KAVACH identifies the braking requirement.

KAVACH communicates braking signal(s) to BIU.

BIU communicates signals to existing brake system of locomotive

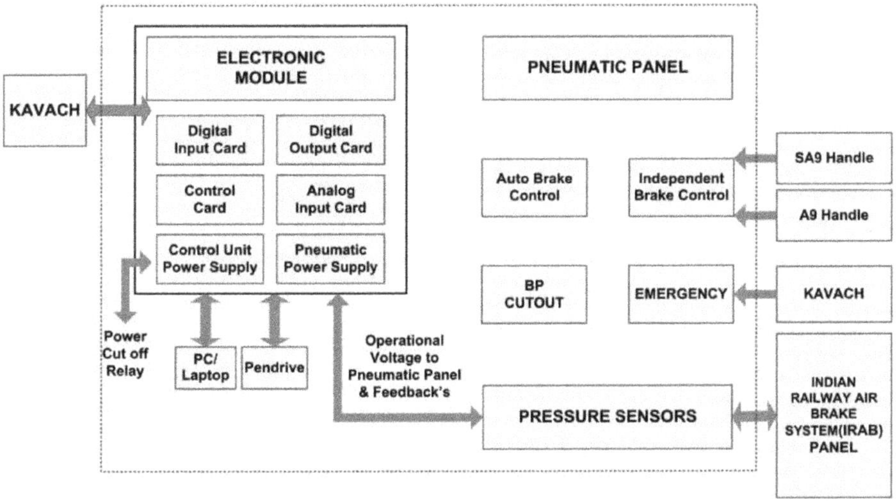

Fig.1.21 Loco Brake Interface

1.21 Block Diagram of Brake Interface Unit

BIU communicates signal for cutting off traction of locomotive.

BIU generates audio-visual indication for Loco Pilot.

BIU gives status of execution of braking command to KAVACH.

1. **Features of BIU:** BIU works parallel to Locomotive Brake System.

   Does not affect brake characteristics of locomotive/train.

   Can override the Loco Pilot Braking and Vice-Versa - Higher braking prevails.

   BIU initiated braking cannot be reduced by Loco Pilot.

   Manual and automatic isolation possible.

2. **Health Monitoring by BIU:** Periodical monitoring of pressures at short interval - Confirms required pressures are maintained.

   If feedback is not correctly received - Automatic Emergency Brake

   Analyzes brake application commands for correct execution – Brake Pressure (BP) drop or build up.

   Heart-beat monitoring of pneumatic valves- voltage and current once every 200ms - Any failure, message alert to LP through KAVACH display and automatic Emergency Brake.

   Power supply 'ON' health monitoring - Alert to LP through KAVACH display.

   Fail safety.

   Loss of electric power leads to Automatic Emergency Brake.

   Loss of communication between Loco KAVACH unit and BIU results in Automatic Emergency Brake.

   For any brake application Traction power is cut-off.

3. **Braking Logic:** The onboard equipment shall have provision for acquiring the braking characteristics through DMI as per the selections made by Loco Pilot at the start of mission or whenever there is change in train consist.

   The brake characteristics shall be such that in the event of perceived danger, the loco unit shall be able to stop train short of safe distance or control the speed to desired value before target. This distance should be possible to be configured during installation with nominal values as 300m for Rear end Collision prevention, stop immediately on detection of Head-on Collision prevention and short of Signal at Danger in case of SPAD prevention.

   By pressing Manual Brake Test (MBT) button by the Loco Pilot in stationary condition of the train, working of all brake valves of Brake Interface Unit (BIU) can be tested.

   The braking logic of the Loco unit shall be so intelligent that based on the (i) brake characteristics of the train, (ii) speed of the train and (iii) gradient of the location & the target, Loco unit shall decide which type(s) of brake and when to be applied to stop the train short of safe distance or control the speed to desired value before target without frequent repeated braking.

   Design of the Loco unit equipment shall be such that its brake interface unit can be isolated by the Loco Pilot as and when required.

   The isolation of brake interface shall be communicated to the Network Monitoring System (NMS) through GPRS/LTE-R.

   Traction cut off feature through KAVACH shall also be isolated under such events.

1.22 Connectivity of Stationary KAVACH Unit with Interlocking

Stationary KAVACH unit shall be capable of taking potential free inputs from interlocking through double cutting arrangement. It shall be capable of taking minimum 256 inputs. (The maximum number of field inputs are 2048). There shall be provision for expansion by providing additional Input cards or Multiple Field Input Units (FIUs). Each Vital Input Card/ FIU can accommodate 256 inputs. Based on the Field inputs required to be connected with each stationary KAVACH units. The requisite number of Vital Input cards/ FIUs shall be procured for each stationary KAVACH.

The break status of potential free contact shall indicate absence of input.

Signal aspect status, position of points, berthing track circuit status, status of track circuits nominated for computing train length and status of block instrument Line Closed condition shall be interfaced to Stationary KAVACH. IBS & Gate unit shall not require inputs for point position, track circuits nominated for computing train length & berthing track circuit status. Gate unit shall not require input for status of block instrument Line Closed condition

1.23 Fall Back Procedures

(a) **Radio communication failures:**A radio communication failure shall be deemed to have occurred when 30 seconds for Absolute Block Section and 10 seconds for Automatic Block Section have passed since the last packet received from Stationary KAVACH in communication mandatory zone.

If the last packet received from Stationary KAVACH unit is more than 6 seconds older, the signal aspect and signal description shall be made blank. However, the Loco KAVACH shall continue to function in Full Supervision mode and shall supervise the Movement Authority received in latest packet.

In the event of a Radio Communication failure longer than applicable time-out, the Loco KAVACH unit shall transit from Full Supervision mode to Limited Supervision Mode and if Loco Pilot does not acknowledge within stipulated time Loco KAVACH unit shall apply service brake. In addition, it shall send the message to Network Monitoring System (NMS) through GPRS/LTE-R.

Stationary KAVACH unit shall send the Fault message to NMS through Ethernet/ GPRS/LTE-R interface.

In the event of a failure of only one radio when other radio is providing radio communication in hot standby, the Loco KAVACH unit should log the fault. Loco KAVACH Unit shall also send the message to NMS through GPRS/LTE-R. Stationary KAVACH unit shall send the Fault message to NMS.

(b) **RFID Reader failures:**If both RFID readers fail, Loco KAVACH unit should stop radio communication and shall switch to System Failure mode. It shall send the message to Network Monitoring System (NMS) through GPRS/LTE-R.

In the event of any one RFID reader failure, Loco KAVACH unit should log the event. In addition, it shall send the message to NMS through GPRS/LTE-R.

(c) **GPS/GNSS failure:**Station/LC/IBS Vital Computer shall have Real Time Clock (RTC) synchronization facility with GPS/ GNSS clock to synchronize with other KAVACH systems in hot standby manner

Incremental difference between the CPU time and GPS time is to be cross checked. If the incremental difference between CPU and GPS time is not matching, the time reference shall change to other GPS. If the difference between two GPS is greater than the frame interval, message shall be sent to Network Monitoting System (NMS).

Diverse make of GPS are preferable to avoid common cause failures.

In the event of failure of both – GPS/GNSS and Real Time Clock (RTC), the Loco KAVACH unit shall stop radio communication and shall switch to System Failure mode. In addition, it shall send the message to NMS through GPRS/ LTE-R.

If the incremental difference is also not matching with second GPS or both systems are failed, then the system shall work on CPU time for 30 minutes, (default, Min: 10, Max: 60) until the situation is stabilized. If there is no stability after GPS time-out, the Loco shall transit out of Full Supervision Mode.

In the event of failure of both – GPS/ GNSS and RTC, Stationary KAVACH unit should stop radio communication and shall switch to System Failure mode. Fault message shall be communicated to NMS through Ethernet interface. In addition, it shall send the message to NMS either through Ethernet or GPRS/LTE-R.

(d) **Driver Machine Interface (DMI) communication failures:**In the event of Active Cab/ Desk DMI communication failure, Loco KAVACH unit shall switch to System failure mode. In addition, it shall send the message to Network Monitoring System (NMS) through GPRS/LTE-R.

In the event of Non-Active Cab/ Desk DMI communication failure, Loco KAVACH unit shall log the fault. In addition, it shall send the message to NMS through GPRS/LTE-R.

1.24 Protection Functions

(a) **Prevention of Signal Passing at Danger (SPAD):** Stationary KAVACH unit shall calculate the movement authority based on the signal Aspect or/and track circuit status, point position and the status of the berthing track circuit.

In case of any conflict between signal aspect, point position, berthing track section, signal aspect sequence and TIN, the Stationary KAVACH unit shall transmit most restrictive aspect of that signal and shall reduce the movement authority accordingly.

Stationary KAVACH unit shall check route information configured on the basis of the KAVACH Control Table of the stationary KAVACH Unit (excluding overlaps).

The off aspect and movement authority for LSS shall be transmitted by Station/ IBS unit only when LSS is off and it is ensured that the concerned Line Clear is available.

(b) **Unusual Stoppage in Block Section:** In case of unusual stopping of train in the block section, if Movement Authority is greater than 300m (configurable), Loco KAVACH unit shall transmit the 'Side Collision' message, after a delay of 15 seconds (configurable) of stoppage of the train unless acknowledged by the Loco Pilot.

(c) **Protection of Roll Back:**Loco KAVACH unit shall be capable of detecting Roll Back of the train through train interface. It shall apply brake and give audio/ visual warning if train has rolled back by more than 5 meters (configurable). KAVACH needs Direction sensing type speed sensor.

(d) **Prevention of Head on & Rear end Collisions in Block Section:**Loco KAVACH units either directly or through Stationary KAVACH unit, shall be capable of detecting head on collisions, rear end collisions of trains/locos on single line, multiple lines in all possible scenarios based on the track identification, speed of the trains, train location, train length, train direction movement (Nominal/ Reverse) etc.

In case of head on collision situation, Loco KAVACH units of both the trains shall automatically apply brakes immediately with warning either in Absolute or in Automatic Block Section.

In case of rear end collision situation, Loco unit of only rear train shall automatically apply brakes to bring it to stop short of stipulated distance (300m in block section, configurable) from the train ahead.

In Station sections, Stationary KAVACH shall prevent train collisions with the help of SPAD and TIN conflict.

In case two trains are detected by Stationary KAVACH to be moving towards each other on same TIN in adjacent block section, the SoS command would be generated by Stationary KAVACH for both the trains. On reception of such Loco specific SoS from Stationary KAVACH Unit, the Trains would be stopped through automatic application of brakes.

In case of Multiple locos, the rear end collision message shall be displayed with the details of approaching loco/train only

(e) **Manual SoS generation/ Cancellation:** Loco as well as Stationary KAVACH unit shall have provision of sending SoS message by pressing SoS and Common buttons together.

Loco as well as Stationary KAVACH unit shall have provision to cancel the Manual SoS message by pressing 'Common' and 'ACK/Cancel' buttons together.

When the SoS and Common buttons are pressed simultaneously in Loco or Stationary KAVACH units, the Loco KAVACH units of all the trains/ Locos within 3000m of Location of SoS originating source as well as self-train (if SoS is generated by Loco unit) and approaching towards Location of SoS originating source, shall apply brakes to bring the train/locomotive to standstill before reaching the originating Location of "SoS" message. After the train speed is reduced to zero kmph, train speed shall be supervised for 30kmph (configurable) till the train passes the originating Location of "SoS" message.

SoS sending as well as receiving KAVACH equipment shall log sending & receiving of SoS message. In addition, the information shall be sent to Network Monitoring System by Loco KAVACH through GPRS/LTE-R and StationaryKAVACH units either through Ethernet or GPRS/LTE-R.

(f) **Train trip:** When a Train in Full Supervision or Limited Supervision Mode passes a stop signal at ON or End of Authority + 30m, the emergency brake shall be triggered.

Operation of the train trip shall be indicated on the DMI.

The emergency brake shall be applied until the Train comes to halt.

When the Train is stationary, the loco pilot shall be required to acknowledge the train trip condition. This acknowledgement will release the emergency brake.

(g) **Auto whistling on approach of Level Crossing Gate:** Auto whistling feature on approach of level crossing gate is optional.

Loco KAVACH unit shall display the level crossing gate information (Gate ID) on DMI, when approach of LC Gate is detected through LC Gate Tags.

Loco KAVACH unit shall blow the Loco horn at LC gate, based on the information received from LC gate tag/ SSP

Loco KAVACH unit shall not blow the horn for LC gate, if movement authority is less than the LC gate distance from its current position.

Continuous whistling shall commence from a distance of 600m on approach of a LC Gate till the time that train reaches LC Gate. Whistling pattern shall be configurable.

1.25 Protection during Transient Conditions

(i) **Radio Communication failure:** If the communication timeout is longer than 6 seconds, the signal aspect and signal description shall be made blank. However, the Loco KAVACH shall continue to function in Full Supervision mode and shall supervise the Movement Authority received in latest packet.

If Radio Communication failure is longer than 30 seconds, the Loco KAVACH unit shall transit from Full Supervision mode to Limited Supervision mode and shall seek acknowledgement from Loco Pilot. If Loco Pilot does not acknowledge within stipulated time of 15 seconds (Configurable), Loco KAVACH unit shall apply service brake. In addition, it shall send the message to Network Monitoring System through GPRS/LTE-R. Stationary KAVACH unit shall send the Fault message to Network Monitoting System through Ethernet/ GPRS/LTE-R interface.

In the event of a failure of only one radio when other radio is providing radio communication in hot standby, the Loco KAVACH unit should log the fault. Loco KAVACH Unit shall also send the message to Network Monitoting System through GPRS/LTE-R. Stationary KAVACH unit shall send the Fault message to Network Monitoting System.

(ii) **Signal goes to danger on approach:** When a stop signal on approach is thrown back to danger, and the train in Full Supervision or Limited Supervision Mode passes the stop signal at ON or End of Authority + 30m, the emergency brake shall be applied until the Train comes to halt.

Operation of the train trip shall be indicated on the DMI.

When the Train is stationary, the loco pilot shall be required to acknowledge the train trip condition. This acknowledgement will release the emergency brake.

After the acknowledgement, the loco pilot shall be able to continue the movement in Post- Trip Mode.

Loco KAVACH unit shall supervise the train against a ceiling speed of Post Trip Mode (Default: 15 kmph) and shall exit the Post Trip Mode after crossing the next approaching signal at OFF.

(iii) **Signal, point or track circuit, failure:** As mentioned earlier, Stationary KAVACH unit is interfaced with station interlocking and calculates the movement authority based on the signal aspect or/and berthing track circuit status or/and route locking status or/and point position. Hence in case of any abrupt failure of signal aspect, point position or berthing track in the face of an approaching train, the movement authority is reduced accordingly and an automatic application of brakes is initiated.

CHAPTER

2 RFID TECHNOLOGY

2.0 Overview

This chapter describes the basic components of a Radio Frequency Identification (RFID) system and explores the technology, applications, and competitive advantages of RFID technology.

2.1 Introduction

RFID offers greater flexibility, higher data storage capacities, increased data collection throughput, and greater immediacy and accuracy of data collection.

An increasing number of companies in a variety of markets worldwide are embracing RFID technology to increase quality and quantity of data collection in an expeditious manner. The technology's enhanced accuracy and security makes it an ideal data collection platform for a variety of markets and applications, including healthcare, pharmaceutical, manufacturing, warehousing, logistics, transportation and retail.

2.2 Components of an RFID System

A basic RFID system consists of these components:

- A programmable RFID tag/inlay for storing item data consisting of:
  - an RFID chip for data storage
  - an antenna to facilitate communication with the RFID chip.
- A reader/antenna system to interrogate the RFID inlay.

2.3 The RFID Tag

RFID tags are categorized as either **passive** or **active**. Passive tags do not have an integrated power source and are powered from the signal carried by the RFID reader. Active tags have a built-in power source, and their behavior can be compared to a beacon. As a result of the built-in battery, active tags can operate at a greater distance and at higher data rates in return for limited life driven by the longevity of the built in battery and higher costs. For a lower cost of implementation, passive tags are a more attractive solution.

The RFID tag consists of an integrated circuit (IC) embedded in a thin film medium. Information stored in the memory of the RFID chip is transmitted by the antenna circuit embedded in the RFID inlay via radio frequencies, to an RFID reader. The performance characteristics of the RFID tag will then be determined by factors such as the type of IC used, the read/write capability, the radio frequency, power settings, environment, etc.

The information stored in an RFID chip is defined by its read/write characteristics. For a **read-only** tag, the information stored must be recorded during the manufacturing process andcannot be typically modified or erased. The data stored normally represents a unique serial number which is used as a reference to lookup more details about a particular item in a host system database. Read-only tags are therefore useful for identifying an object, much like the "license plate" of a car.

For a **read/write** tag, data can be written and erased on demand at the point of application. Since a rewriteable tag can be updated numerous times, its reusability can help to reduce the number of tags that need to be purchased and add greater flexibility and intelligence to the application. Additionally, data can be added as the item moves through the supply chain, providing better traceability and updated information. Advanced features also include locking, encryption and disabling the RFID tag.

RFID systems are designed to operate at a number of designated frequencies, depending on the application requirements and local radio-frequency regulations:

Low Frequency (125KHz)

High Frequency (13.56MHz)

Ultra High Frequency (860-960MHz)

Microwave (2.45GHz)

Low-frequency tags are typically used for access control & security, manufacturing processes, harsh environments, and animal identification applications in a variety of industries which require short read ranges. The low frequency spectrum is the most adaptive to high metal content environments, although with some loss of performance. Read ranges are typically several inches to several feet.

High-frequency tags were developed as a low cost, small profile alternative to low-frequency RFID tags with the ability to be printed or embedded in substrates such as paper. Metal presents interference issues and requires special considerations for mounting. Similarly to the low-frequency technology, these tags have a read range of up to several feet.

UHF tags boast greater read distances and superior anti-collision capabilities, increasing the ability to identify a larger number of tags in the field at a given time. The primary application envisioned for UHF tags is supply chain tracking. There are large number of additional markets with demand for UHF RFID technology such as transportation, healthcare, aerospace, etc.

ATP-KAVACH RFID Tags operate in UHF band (865 to 867 MHz)

Microwave tags are mostly used in active RFID systems. Offering long range and high data transfer speeds at significantly higher cost per tag making them suitable for railroad car tracking, container tracking, and automated toll collection type applications as a re-usable asset.

The table highlights the Max Read Range of the RFID tags for the different operating frequency ranges:

Frequency Range	125 KHz	13.56 MHz	868 - 956 MHz	2.45 GHz & 5.8 GHz
Typical Max Read	Low Frequency (LF)	High Frequency (HF)	Ultra-High Frequency (UHF)	Microwave
Range (Passive Tags)	< 0.5 m	~1 m	~3m to ~10m	~1 m

2.4 Regulations in INDIA

The Department of Telecommunication (under the Ministry of Communications and Information Technology), has delicensed the spectrum in the 865–867 MHz band for use by RFID devices. The regulation on the use of wireless equipment in the band 865–867 MHz specifies that no license is required by any person to establish, maintain, work, possess, or deal in RFID on non-interference, non-protection, and nonexclusive bases, in the frequency band 865–867 MHz with maximum 1 W transmitter power, 4 W ERP, and 200 kHz carrier bandwidth.

2.5 RFID Summary

The dominant RFID dynamic behind supply chain applications is the EPC standard using the UHF frequency band: 902-928 MHz (North America) and 865-868 MHz (Europe). EPC Global, a joint venture between GS1, Inc. (formerly EAN International) and GS1 US (formerly the Uniform Code Council [UCC]) is focused on helping supply chains and industry implement the Electronic Product Code™ (EPC) through the development of global standards and support of the EPC global network™. The EPC Global Network ideally intends to transform the global supply chain through a new, open global standard for real-time, automatic identification of items in the supply chain of any company, in any industry, anywhere in the world.

2.6 Introduction to RFID Tags

This chapter describes the RFID fixing arrangement guidelines to be ensured while designing the RFID fixing arrangement for the purpose of KAVACH System. This is based on Joint Procedure Order approved by Track and Signal Directorates of RDSO (STS/E/KAVACH/Tender/Part-VIII dated 27.08.2015).

2.7 RFID Tag Used in KAVACH

RFID Tags provide location specific static information to Locomotive. Apart from acting as Location references these provide immediate information such as crossing the signal etc. to Loco Unit.

RFID tags shall be fitted on track in station section, point zones, near Signals & in block section for giving Trackside information to Loco KAVACH unit.

The RFID tags shall be fitted on the sleepers between the rails as per guidelines given for Indian Railways.

Specification of RFID tags:

(a) Suitable for reliable working at train speed upto 200 KMPH (minimum).

(b) Frequency of operation: 865-867MHz.

(c) Can be programmable with minimum 128 bits (including CRC) of user data.

(d) Shall be able to work even when submerged in water up to rail level.

(e) Under field operating conditions RFID reader antenna shall be able to read,

RFID tag from a vertical distance of 700 mm from bottom of RFID reader antenna to top of the rail level.

Fig. 2.1

2.8 Preparation of KAVACH RFID Tag –TIN Layout

RFID tag-TIN layout shall be prepared as per guidelines given in RDSO Draft Specification no. RDSO/SPN/196/2012 Ver 3.2 or latest. For preparation of Tag-TIN Layout, one should have a fair idea about classification of RFID Tags and Track Identification Numbers (TIN). The details are given in following paragraphs:

2.9 Classification of RFID Tags

RFID tags are categorized as follows:

(a) Normal tag

(b) Signal foot tag

(c) Signal approach tag

(d) Track Identification Number (TIN) Discrimination tag

(e) LC gate tag(optional)

(f) Tunnel tag (for future use)

(g) KAVACH Exit tag

2.10 Following are the Guidelines for provision of RFID Tags in Tag-TIN Layout

(a) Normal tags shall be provided in the block section as well as in station section. The maximum distance between the two normal tags shall not be more than 1000m. Each Normal tag shall be linked to next two normal tags in both the directions (Nominal & Reverse).

(b) Signal foot tags shall be provided at foot of every signal post.

(c) Signal approach tags shall be provided before the approach of (typically 150 ~ 250m) every signal post to correct the odometry error.

(d) TIN Discrimination tags shall be used to indicate change in the TIN of track section.

(e) Normally it will be placed at turnouts.

(f) LC gate tag (optional) shall be provided at both sides of LC gate as required by operating Railway.

(g) KAVACH Exit tag shall be provided at KAVACH territory exit point.

Normal Tag:

Field Description	Data to be configured
TAG ID	RFID Tag ID as per Layout Range: 1 to 1023
Absolute Loc (in Meters)	Absolute Location in meters as per Layout. Range: 0 to 2621420 m
TIN in Nominal Dir	TIN Number as per Layout in Nominal Direction. Range: 0 to 127
TIN in Reverse Dir	TIN Number as per Layout in Reverse Direction. Range: 0 to 127

Contd...

Field Description	Data to be configured
Trans from Loco-to-Stationary is necessary in Nominal Dir	Drop down menu: Yes or No
Trans from Loco-to-Stationary is necessary in Reverse Dir	Drop down menu: Yes or No
Station/ Block ahead info in Nominal Dir	Drop down menu: Station or Block section
Station/ Block ahead info in Reverse Dir	Drop down menu: Station or Block section
Reserved	Non-configurable and default value is 0

Signal Foot Tag.:

Field Description	Data to be configured
TAG ID	RFID Tag ID as per Layout Range: 1 to 1023
Absolute Loc (in Meters)	Absolute Location in meters as per Layout. Range: 0 to 2621420m
TIN in Nominal Dir	TIN Number as per Layout in Nominal Direction. Range: 0 to 127
TIN in Reverse Dir	TIN Number as per Layout in Reverse Direction. Range: 0 to 127
Station ID	Station ID as per Layout. Range: 1 to 1023
Signal Direction	Drop down menu: Nominal or Reverse

Signal Approach Tag. :

Field Description	Data to be configured
TAG ID	RFID Tag ID as per Layout Range: 1 to 1023
Absolute Loc (in Meters)	Absolute Location in meters as per Layout. Range: 0 to 2621420m
TIN in Nominal Dir	TIN Number as per Layout in Nominal Direction. Range: 0 to 127

Contd...

Tin Discrimination Tag:

Field Description	Data to be configured
TAG ID	RFID Tag ID as per Layout Range: 1 to 1023
Absolute Loc (in Meters)	Absolute Location in meters as per Layout. Range: 0 to 2621420m
TIN in Nominal Direction	TIN Number as per Layout in Nominal Direction. Range: 0 to 127
TIN in Reverse Direction	TIN Number as per Layout in Reverse Direction. Range: 0 to 127
Station ID	Station ID as per Layout. Range: 1 to 1023
Dead end ahead in Nominal Direction	Drop down menu: Yes or No
Distance of dead-end in Nominal Direction	If Yes, distance to dead end in meters as per Layout. Range: 0 to 255 in decameters
Dead end ahead in Reverse Direction	Drop down menu: Yes or No
Distance of dead-end in Reverse Direction	If Yes, distance to dead end in meters as per Layout. Range: 0 to 255 in decameters

LC Gate Tag.:

Field Description	Data to be configured
TAG ID	RFID Tag ID as per Layout Range: 1 to 1023
Absolute Loc (in Meters)	Absolute Location in meters as per Layout. Range: 0 to 2621420m
LC Gate Approach Tag in direction of applicability	Drop down menu: KAVACH Non-KAVACH First Non-KAVACH Second Spare
LC Gate in Nominal direction	Drop down menu: Yes or No
Gate ID	Gate ID as per Working Time Table Range is from 1 to 1023

KAVACH Exit Tag.:

Field Description	Data to be configured
TAG ID	RFID Tag ID as per Layout. Range: 1 to 1023
Absolute Loc (in Meters)	Absolute Location in meters as per Layout. Range: 0 to 2621420 m
Applicable Direction	Drop down menu: Nominal or Reverse

2.11 Track Identification Number (TIN)

(a) Each track shall have designated Track Identification Number (TIN).

(b) Each Block section shall have single unique designated TIN. Block Section TIN can be repeated after a designated distance (50 km minimum along the track route).

(c) To avoid unnecessary SOS generation, adjacent TINs to be incorporated in the radio packet and adjacent line tag info. Also, location adjustment details shall be sent on radio only after there is physical separation between the tracks.

(d) Each line in the station section having berthing portion shall have different TINs. TIN can be repeated after a designated distance (10 km minimum along the track route).

(e) TINs shall be allotted in such a manner not to inhibit permissible simultaneous movements.

(f) Loco KAVACH unit shall be able to self-deduce the change in its TIN whenever it changes the TIN section.

Numbering Scheme for Stationary KAVACH: The first two digits are allotted to each zonal railway and Zonal railways may decide the last three digits for the stations. A proper record shall be maintained by the Zonal Railways to avoid repetition of number as station ID should be unique number.

2.12 Guidelines for preparation of RFID TAG-TIN Layout

Following guidelines shall be followed while preparing RFID tag-TIN layouts for Station/IB/LC or block sections:

1. RFID tag-TIN layout shall be prepared with Station yard layout as reference. However, the actual site considerations shall be taken into account prior to its preparation. A site survey shall be conducted to mark the locations where tags need to be placed.
2. The centre of Station Master's panel shall be taken as station's Centre Line for reference purpose.

3. Normal tags shall be provided in the block section as well as in station section. The maximum distance between the two Normal tags shall not be more than 1000m.
4. Every signal, including shunt signals, shall be provided with Signal foot tag.
5. The permanent EoA (End of Authority) Locations such as Stop Boards, Block section limiting board (BSLB), Ends of berthing track (may be with shunt signal sometimes) shall have Signal foot tag.
6. Signal Approach tag shall be provided for every Signal foot tag. They shall be provided at a distance of 150-250m from Signal foot tag. The distance from Signal Approach tag to Signal shall be mentioned on the layout.
7. All type of tags, except Junction and Adjustment tags, can be placed instead of Signal Approach tag if it offers some other advantage like reduction in number of tags.
8. To ensure linking up to the nearest location to the approaching signal, Normal tag shall be provided, in lieu of signal approach tag on Mainlines.
9. To demarcate TIN sections, TIN discrimination tags shall be placed.
10. Gate tags shall be placed at such a distance that Auto whistling for approaching LC gate can commence from at least 600m on approach of LC gate.
11. Adjustment tag shall be used, if possible, in a non-signalling area. The layout shall only mention the programmed absolute location. However, the physical distance between the Adjustment tag and nearest tag, from where physical and programmed distances are different, shall be mentioned.
12. Junction tag shall be used, if possible, in a non-signalling area. The layout shall only mention the programmed absolute location. However, the physical distance between the Junction tag and nearest tag, from where physical and programmed distances are different, shall be mentioned.
13. While moving from KAVACH to non KAVACH territory, Exit tags shall be provided at the exit boundary of Last Stationary KAVACH unit in KAVACH territory. Exit tags shall be provided in 3 sets of tags.
14. The distance between two different type of tags shall not be less than 40m. However, in case of Exit tags, the same may be reduced to 20m.
15. The distance between duplicated tags shall not be more than 2 meter.
16. A single TIN section shall be represented using a single colour. The TINs in vicinity shall be represented in different colours.
17. Non-KAVACH territory shall be represented through white colour.
18. The TIN layouts thus prepared, shall permit all the train movements allowed in a section as per Table of Control/Selection Table.
19. At all places, where the train is likely to move outside KAVACH territory or remain stabilized for long duration for e.g., sidings, Exit tags shall be provided.

20. Following RFID Tag notations shall be used to denote different types of tags:

Type of Tag	Notation
Normal	N
Signal Foot	S
Signal Approach	SA
TIN Discrimination	T
Gate	G
Exit	X

Fig. 2.2

21. Whenever, exit tags are provided on lines permitting movement in both directions, two Normal tags shall be provided to allow the train to enter in Full Supervision as soon as possible.
22. Block Section TIN should be extended up to Block section limiting board (BSLB) on unidirectional lines (such as Double Line) and up to opposite direction Advanced Starter on Single Line. S-tag shall be provided at the Yard Exit Points not protected by Signals like BLOCK SECTION LIMITING BOARD (BSLB) etc.
23. The distances from a Normal Tag to all other Normal tags ahead in the direction of movement must be same.
24. The distance of Signal Approach (SA) Tag from corresponding Signal Foot (S) Tag should be exact multiple of 10m (1 Decameter). This should be invariably ensured during installation and verification at site.
25. Tags in Block Section shall be placed with consideration to the ease of maintenance and vandalism. These should be placed, if feasible, in vicinity of EC sockets, LC gate or any other place where a maintainer is usually required to visit.
26. Reference drawing numbers shall be mentioned on the layout.

27. Tag numbers (values in the range of 1 – 1023) and TIN numbers (values in the range of 1-127) shall be allotted by the user Railways. Sufficient spares for future needs shall be taken into consideration while allotting the numbers. The allotted numbers shall also be mentioned on the RFID tag-TIN layout.
28. Signature Block and revision history blocks shall be prepared as per the practices of User Railways.
29. Legends mentioning the notation used for the purpose of preparation of layout shall be specifically mentioned on the layout.
30. RFID tag-TIN layout need not be up to scale. If the layout is not per scale, the same shall be mentioned on the layout.
31. Absolute locations of tags, LC gates, signals and turnout switches shall be mentioned on the RFID layout.
32. Absolute location of Station center line shall be mentioned on the layout.

2.13 Preparation of KAVACH Control Table

Following guidelines shall be followed while preparing KAVACH Control table:

1. KAVACH control tables shall be based on the SIP of the station as well as approved RFID tag-TIN layout for the Station/IB/LC. However, overlap points shall not be proven in KAVACH control table.
2. Shunt signals shall not be a part KAVACH control tables. However, station shunt limits shall be specified in the Stationary KAVACH application data.
3. KAVACH control table shall include all signals which will be monitored by a specific stationary KAVACH unit.
4. In case of permissive signals, where the inputs for signal indications are available, the ECR shall be used for the purpose of displaying signal aspect. However, movement authority shall be decided based on the signal aspect of the approaching Stop Signal.
5. In case of permissive signals, where the inputs for signal indications are not available, the signal aspect and movement authority shall be derived based on the signal aspect of approaching stop signal.
6. Following information shall be included as part of control tables:
7. Entry Signal: This shall be the approaching signal for a route.
8. Exit Signal: This shall be the next approaching signal on route.
9. Line: This shall describe the line for route for e.g. Down Main, Common Loop, etc.
10. Aspect of Entry Signal: This field shall indicate all the permissible OFF aspects of the "Entry Signal".

11. Requires Aspects of Exit Signal: This field shall indicate requirement of signal aspect for exit signal for the corresponding "Aspect of Entry Signal".
12. Requires Points in Route: This field shall indicate the point positions required in Normal and Reverse positions for the corresponding route. Overlap points shall not be included for the purpose of KAVACH control table.
13. Requires Track Circuit Up in Route: This field shall indicate the track circuits required to be in picked UP condition for the signal to be taken OFF. Only platform berthing track circuits shall be included for the purpose of KAVACH control table.
14. TINs (Track Identification Number) Requires Free: This shall mention all the TINs falling into the respective route as mentioned in approved RFID layout. TINs shall not be proved for calling on signals.

2.14 Check RFID

Entry Signal Foot Tag: This shall indicate the signal foot tag for the Entry Signal.

En-Route Tags: This shall indicate all the RFID tags falling in the corresponding route as per approved RFID tag-TIN layout. The Signal foot tag for the "Exit Signal" shall not be included in this field.

Conflicting Route Tags: This field shall indicate all the tags on which it is possible for the train to move if the point positions are not in accordance with "Requires Points in Route". For the unintended route in the same direction, this shall include tags up to the last tag on the unintended route. Signal foot tag of next signal and turnout tags shall not be included in the conflicting route tags. For the unintended route in opposite direction, this shall include tags up to and including the Signal foot tag for next signal. If a conflicting route tag is read by a train, the KAVACH system shall generate SoS.

Conflicting turnout tags: The conflicting turnout tags shall be indicated in the KAVACH Control table.

2.15 Turnout

Speed in kmph: Permissible speed for the turnout in route. For the purpose of KAVACH control table, these have been specified as 30 kmph in case of single turnout and 15 kmph in case of multiple turnouts in route. In case, the turnout portion and loop line have different permissible speeds, the speed for both the portion shall be mentioned on the control table.

Distance to Commence: This shall be the distance from foot of "Entry Signal" to the start of first diverging point in route and shall be specified in meters.

Speed Restriction Distance: This shall be the distance from first diverging point in route. In case of entry in to the station premises, this distance shall be the distance from start of first diverging point in route to the next approaching signal plus 30m. In case of exit from the station premises, this distance shall be the distance from foot of approaching signal in route to the end of last converging point plus 10m. In case, the turnout portion and loop line have different permissible speeds, the speed restriction distance for both the portions shall be mentioned on the control table.

Distance between Entry & Exit Signal: To be specified in meters.

Movement Authority: Minimum movement authority for the corresponding signal aspect to be mentioned in meters.

Action for generating SoS to prevent Head on Collision and Rear End Collision on received communication for Block Section TIN: Stationary KAVACH unit shall generate SoS to prevent Head on collision and Rear End Collision in the block section if it receives the required information, even if the train is outside the Stationary KAVACH unit communication boundary.

Entry Signal for Block Section TIN: This shall specify last stop signals in UP/DN direction to be monitored by Stationary KAVACH unit.

RFID Tags: These shall specify all RFID tags on the Block Section TIN for corresponding last stop signal as per RFID tag-TIN layout.

TIN: Block section TIN for the corresponding Last Stop Signal.

RFID TAG-TIN LAYOUT PREPARATION GUIDELINES

The following are the guidelines to be followed in preparation of RFID Tag-TIN Layout for Stations/IB/LC or Block Sections

Study of SIP (Signal Interlocking Plan): The SIP for a given station/yard is the primary input for preparing the RFID Tag – TIN Layout. One has to carefully study the SIP, observe the signals, number of lines, points, berthing tracks, sidings, single ended points, Stop boards, Block section limit boards, Non-signaling zones if any.

Site Survey: After carefully observing the SIP, one has to conduct a site survey which involves visual inspection of the site conditions. The geographical location of centre of the SM panel located in the Station Master room is considered as the reference point for marking the locations of various tags that are placed in the station and its vicinity. From the reference location, the distance of all the signals (including shunt signals) upto last Distant signal on one end and the Advanced starter on the other end, centre of the points, Stop boards, Block section limit boards should be measured. In lines, where there is only one signal on the berthing track, distance of the other end of the berthing track from the reference point should be measured. In addition to this, the distance of various Turn_out switches from the reference point is to be measured which shall be indicated on the Layout.

Allocation of TINs: Each berthing track in the station shall have an unique TIN (Track Identification Number). They shall have to be indicated on the Layout using different colours.

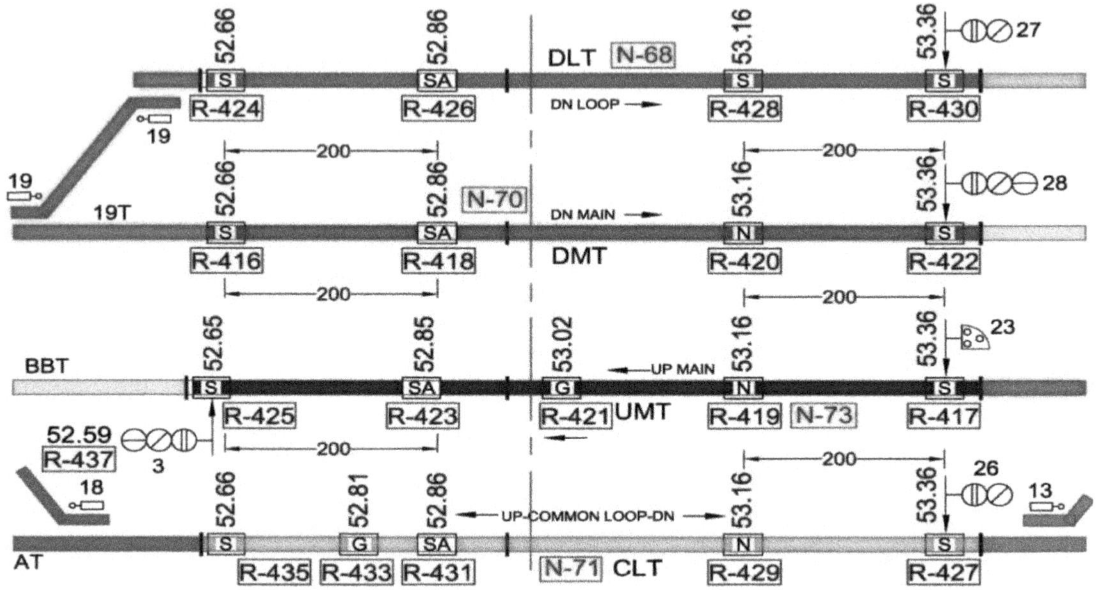

Fig. 2.3 Berthing TIN

Case I: TINs should be allocated in such a way that they do not inhibit permissible simultaneous movements

TIN demarcation to be done at points joining the two Main lines:

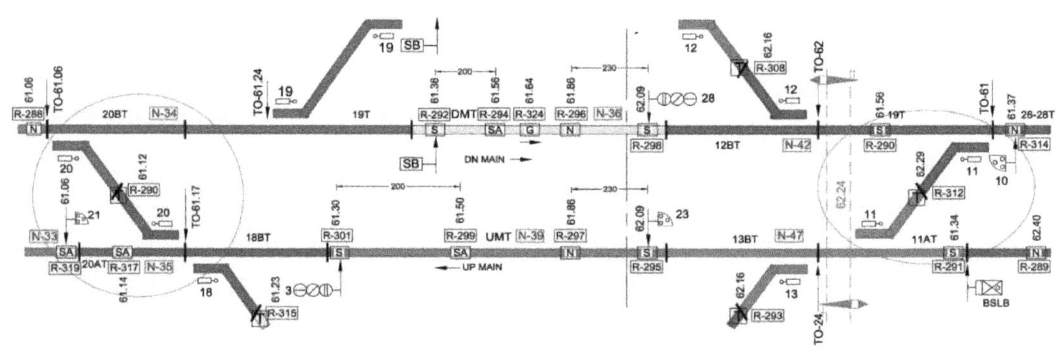

Fig. 2.4

TIN demarcation to be done at points joining the Main line and the Common Loop line:

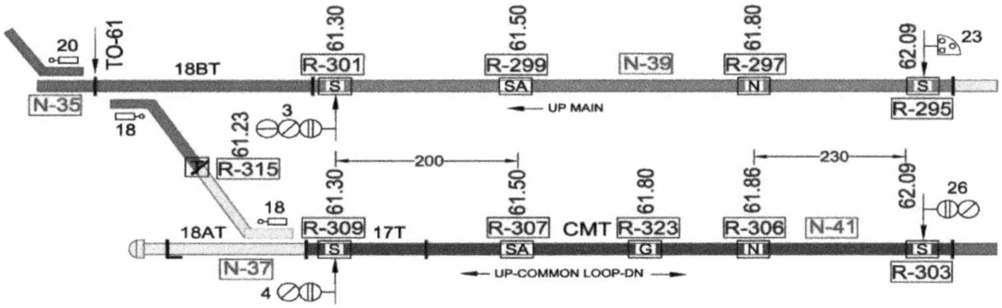

Fig. 2.5

No TIN demarcation to be done at points joining the Main Line and the corresponding Loop line:

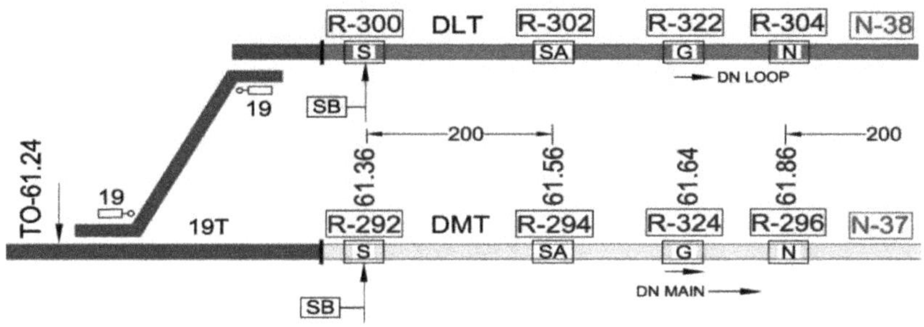

Fig. 2.6

The portion of the track leading to a dead end (sand hump) shall have different TIN:

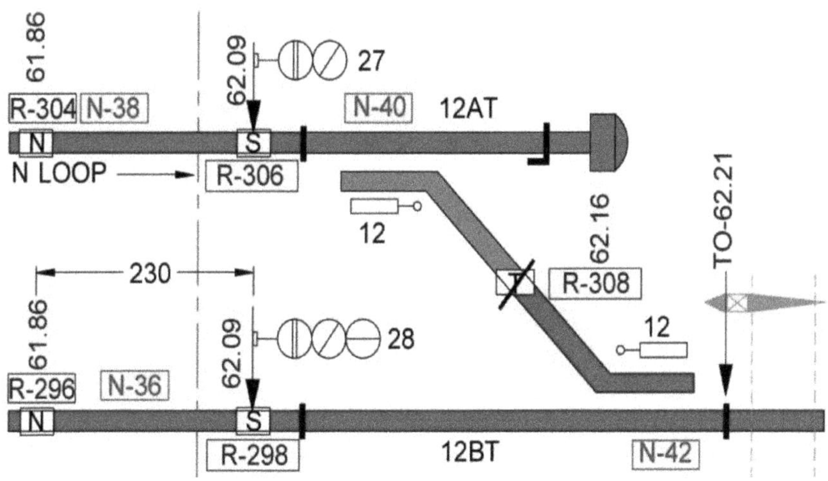

Fig. 2.7

In case of adjacent Common Loop lines, for all the points converging at a single point with just a single sand hump or no sand hump same TIN should be assigned for all the points except the last point joining the Common Loop line to the Main Line:

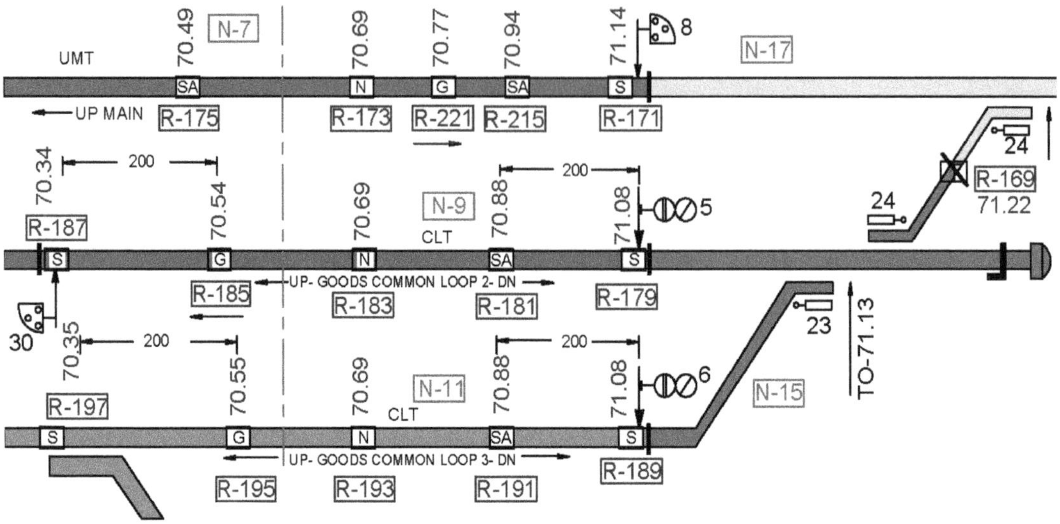

Fig. 2.8

If the adjacent Common Loop lines are accompanied with sand hump for each line, each sand hump shall be assigned with different TIN:

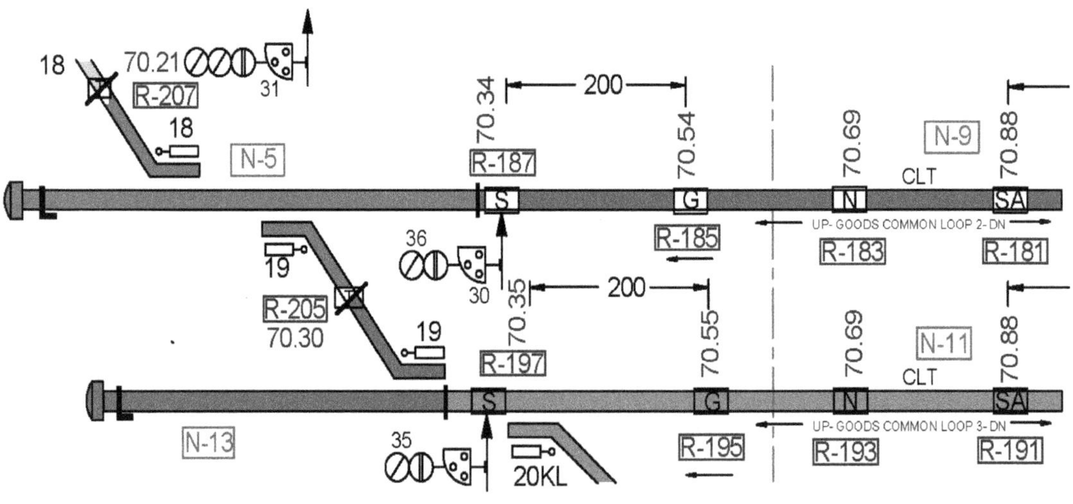

Fig. 2.9

The portion of track between the Shunt signal and the advanced starter shall be assigned a different TIN to allow shunting and movement through adjacent Common Loop line simultaneously:

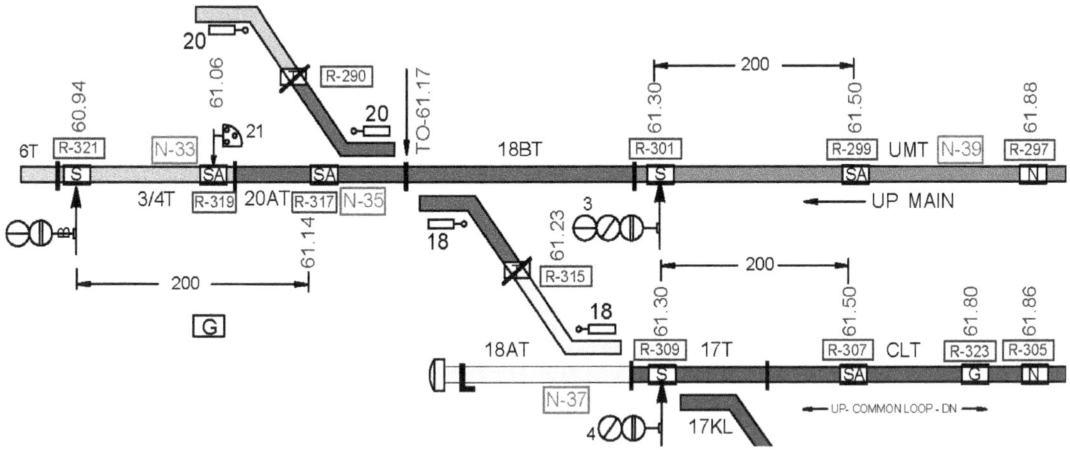

Fig. 2.10

Case II where parallel movement is possible, TINs shall be assigned in such a way that movement in one line do not restrict movement along the other lines:

The following example discusses various cases where TIN assignment is done to allow safe parallel movements. Consider the following sample Layout:

1. **TIN N-2 and N-16:** These are assigned in such a way to allow Train movement from S-17 or S 18 to S-1DA\20 and from S-48 to S-18 or S-19 or S-20 simultaneously. Also, this allows Train movement from S-1DA\20 to S-16 and from S-48 to S-18 or S-19 or S-20 simultaneously
2. **N-25:** TIN is assigned in such a way to allow Train movement from S-18 to S1DA\20 and from S-48A to S-18 or S-19 or S-20 simultaneously
3. **N-27:** TIN is assigned in such a way to allow Train movement from S-18 to S1DA\20 and from S-17 to S-21 simultaneously
4. **N-5:** TIN is assigned in such a way to allow Train movement from S-17 to S1DA\20 and from S- 4 or S-15 or S-16 to S-21 simultaneously.

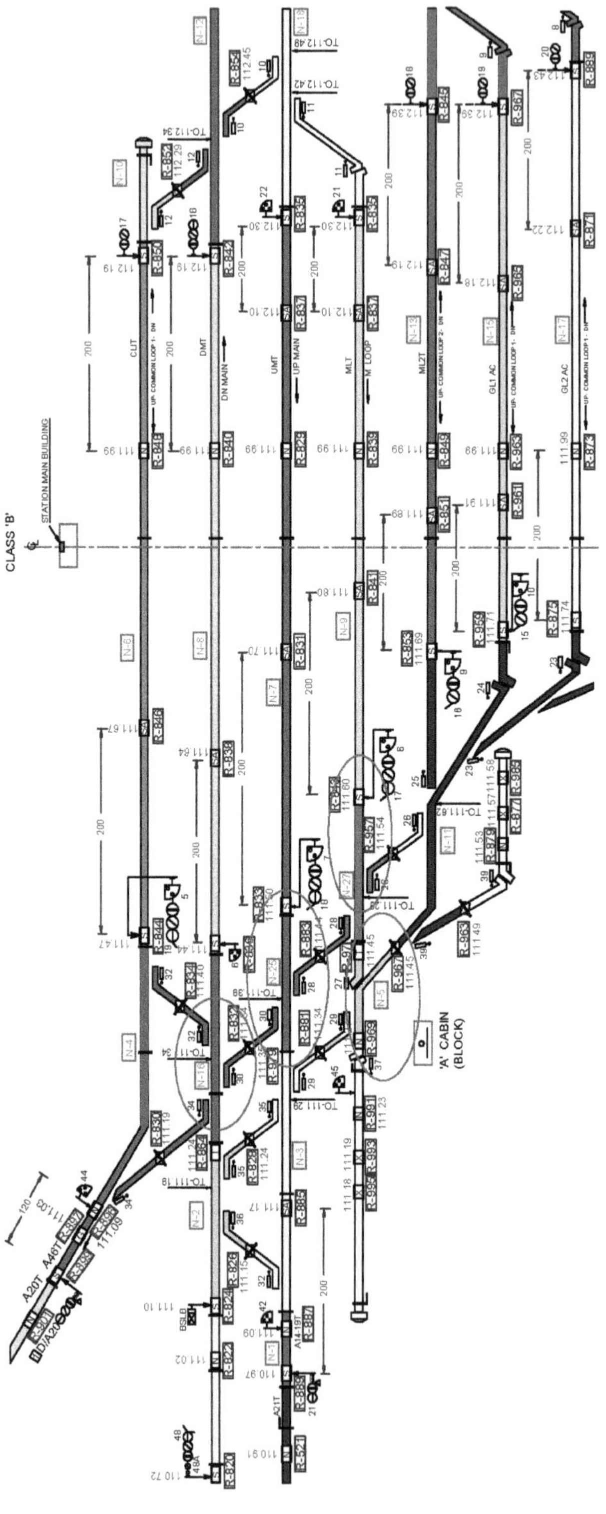
CLASS 'B'
STATION MAIN BUILDING
'A' CABIN (BLOCK)
CUT
DMT
DN MAIN
UMT
UP MAIN
MLT
M LOOP
ML2T
GL1 AC
GL2 AC

Fig. 2.11

In cases where Exit tags are provided with two sets of Normal tags, TIN shall not be assigned upto the second Normal tag which marks the entry into KAVACH territory:

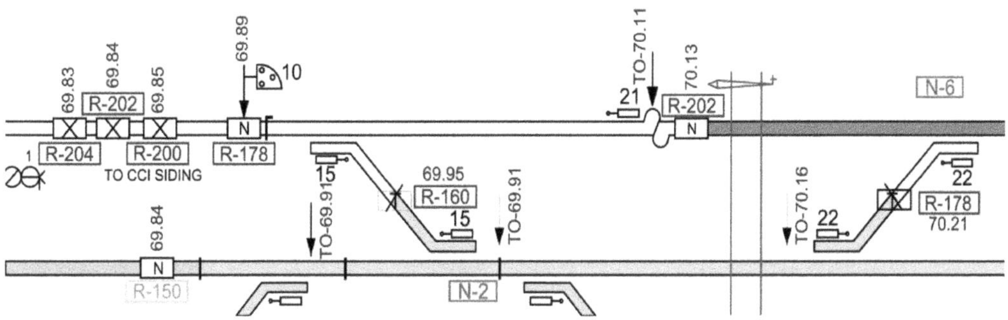

Fig. 2.12

- Each Block section shall have an unique TIN. Block section TIN can be repeated after a designated distance (50 km along the track).
- Block section TIN shall be extended up to BLOCK SECTION LIMITING BOARD (BSLB) on Double line and upto opposite direction Advanced starter in Single line.
- TINs used in one station can be reused in other stations.
- In Non-KAVACH sections or Non-Signalling sections, TINs shall not be provided and shall be indicated in White or transparent colour.

Placement of Tags:

- Normal tags shall be provided throughout the Block section and also within the Station. The distance between two Normal tags should not exceed 1000 metres.
- Normal tags shall be placed in such a way that all the Normal tags ahead in the same direction should be at equal distance.

In cases where, the distance from Normal tags of same absolute location placed in multiple Loop lines (including Common Loop line) to the next Normal tag is varying by a large margin (>50 metres), additional Normal tag should be placed on the same line to reduce the error in distance.

It shall be ensured that a Normal Tag is placed at a distance of 150m to 200m after Advance Starter in Double line and at a distance of 100m to 150m after Advance Starter in Single line which notifies the Loco KAVACH about entry into Block section. This shall make Loco KAVACH to initiate Loco to Loco communication which is useful in prevention of occurrence of any hazardous situations

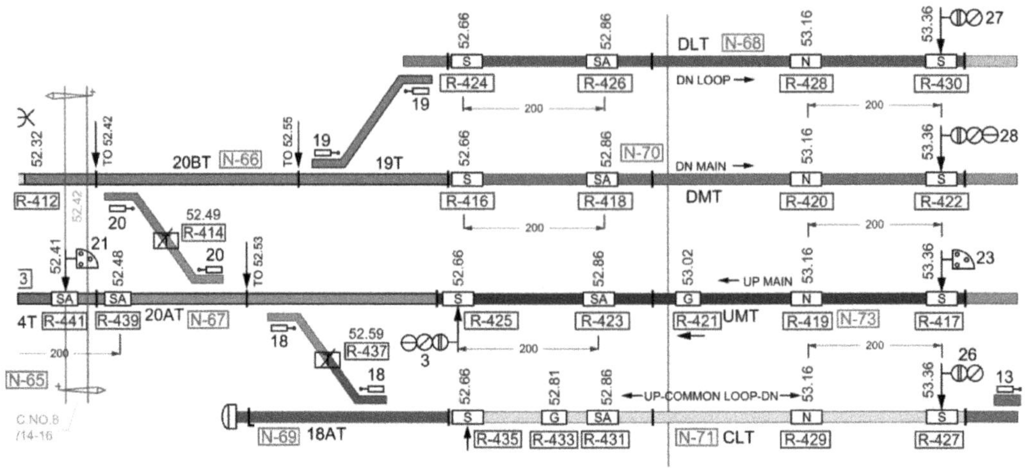

Fig. 2.13

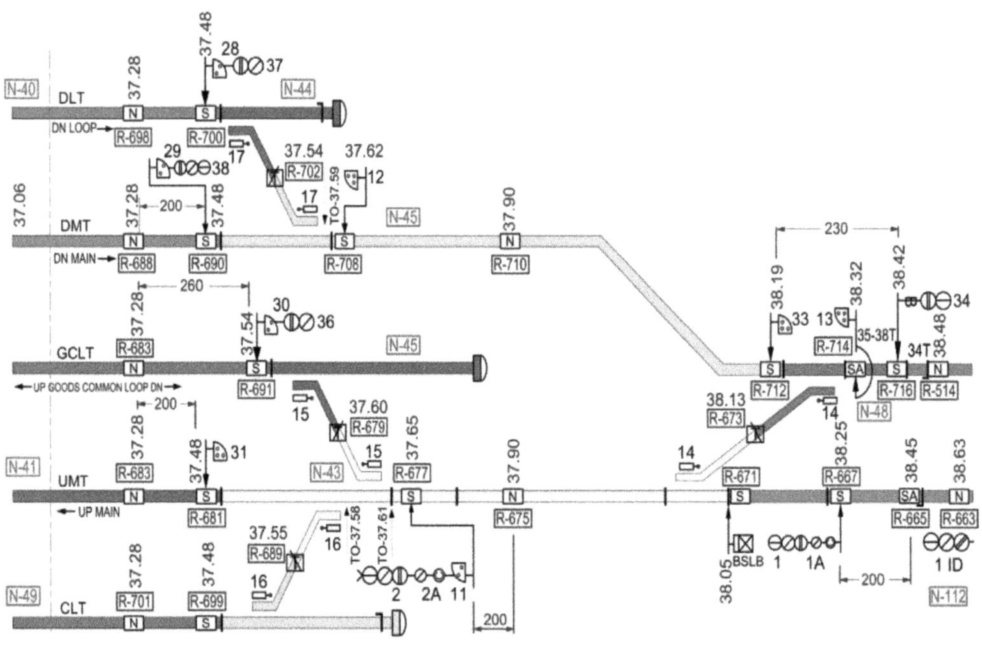

Fig. 2.14

Every signal except shunt signal is to be provided with Signal foot tag. In addition to the signals, the permanent End of authority locations like Stop board, Block section limit board and end of berthing tracks (usually provided with shunt signals) are to be provided with Signal foot tag.

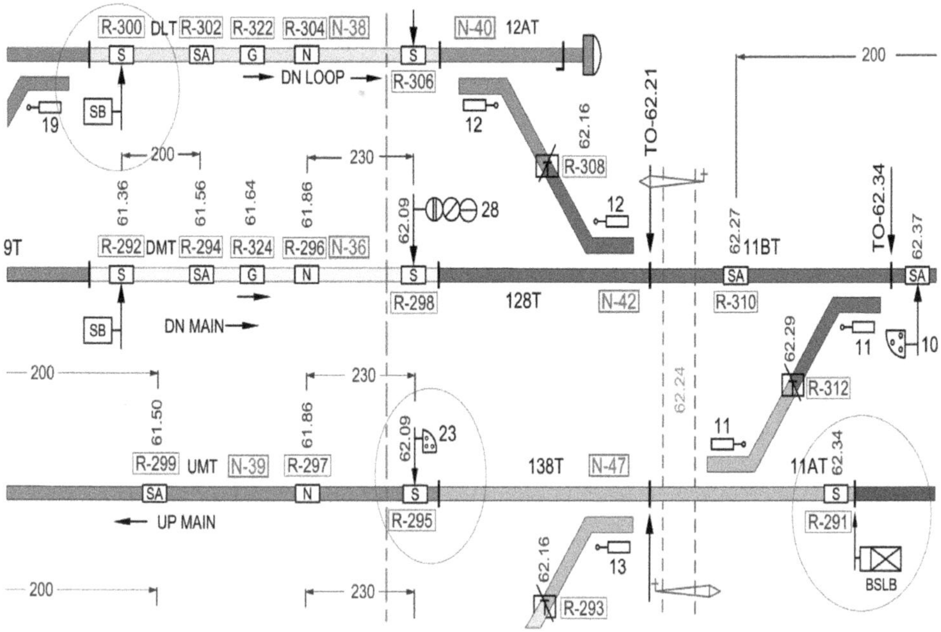

Fig. 2.15

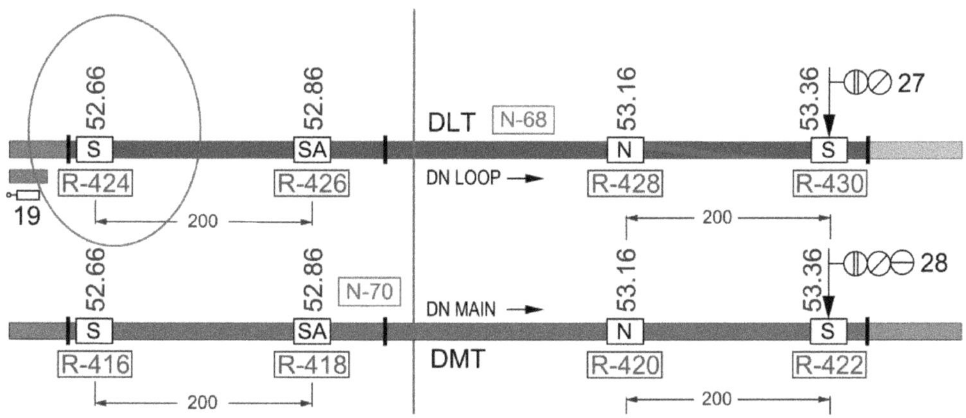

Fig. 2.16

- Shunt signals are to be provided with either Signal approach or Gate tag or Normal tag or Signal foot tag based on fulfillment of all other requirements mentioned in this document.
- Signal Approach tag is to be provided for every Signal and is to be placed at a distance that can vary from 150 metres to 250 metres from the approaching signal. Signal approach tag is not necessary for Signal foot tags placed at BLOCK SECTION LIMITING BOARD (BSLB) and Shunt signals

- All types of tags except Junction tag and Adjustment tag can be used as Signal approach tag if there is an added advantage of reduction in number of tags.
- To differentiate TINs, TIN discrimination tags are to be used.
- Gate tags shall be placed at a distance of 600 meters on main line and 400 meters on loop lines from the approaching LC Gate.
- Adjustment tag to be placed strictly in Non-Signalling area to correct any discrepancy in absolute location in cases where the physical distance between two adjacent stations is not matching the difference between the absolute location of those stations. The physical distance from the adjustment tag to the nearest tag from where the absolute location adjustment is made is to be mentioned on the layout.

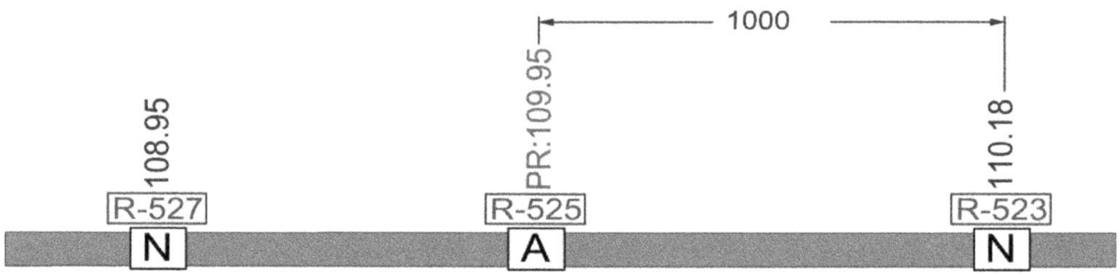

Fig. 2.17

- In sections where there is a change in absolute locations of the Stations usually occurring at Junctions, two sets of Junction tags are to be placed in the Non-signalling area. These are to be accompanied with two sets of Normal tags on either side placed at a distance of 50 meters and 100 meters respectively from the nearest Junction tag. Also, these distances are to be mentioned on the Layout. The absolute location of Junction tag is programmed '0'.
- Exit tags are to be placed where there is a transition from KAVACH territory to Non-KAVACH territory. These shall be provided in three sets of tags. Also in cases where there is a possibility of Trains getting halted for indeterminate time or move out of KAVACH territory, Exit tags are to be provided.
- Whenever Exit tags are placed in places where there is possibility of movement in either directions, two Normal tags are to be placed to enable the KAVACH equipped Train to enter into Full supervision mode on reading these two Normal tags.

- The TIN has to be shown only up to 1st N Tag and TIN shall not be shown after this leading into KAVACH territory exit.
- The distance between any two tags shall be atleast 40 meters but the Exit tags can be placed closely at a distance of upto 20 meters.
- The distance of Signal approach tags from the approaching signal should be in multiples of decameters. This should be strictly ensured and verified at site.
- In the Block section except Signal foot tags, taking into consideration, easy maintenance and to avoid vandalism, tags are to be placed, if feasible, near EC sockets and LC Gates. Make sure that Tags are placed near OHE poles in RE area.
- RFID Tag Ids and TINs shall not be repeated within a station and its corresponding block sections.

Nomenclature to be used:

- Reference drawing numbers shall be mentioned on the Layout.
- Tag numbers (1 to 1023) and TINs (0 to 127) shall be allotted by the user Railways. Sufficient spares shall be reserved for future use. Also, the allotted Tag Ids and TINs should be mentioned on the RFID Tag-TIN Layout.
- Absolute location of Station centre shall be indicated on the Tag Layout.
- Absolute location of Tags, LC Gates, Signals, Shunt Limits, PSR (Permanent Speed Restriction) start/end points and Turn_out switches shall be mentioned on the Layout. This data shall have to be collected from the field survey.
- The distance between the Signal approach tag and the approaching signal is to be mentioned on the Layout.
- Track circuits that include Berthing tracks, track circuits used to calculate Train length, Points and others that are taken as inputs to the stationary KAVACH shall be indicated on the Layout.
- Signature Block and revision history blocks shall be prepared as per the practices of User Railways.
- Legends mentioning the notation used for the purpose of preparation of Layout shall be specifically mentioned on the Layout.
- RFID tag-TIN Layout need not be up to scale. If the Layout is not per scale, the same shall be mentioned on the Layout.

Field	Signal Foot Tag		Description
	Bit positions	No. of bits	
Type of Tag (0001 : Signal Foot Tag)	x3 - x0	4	This field denotes the type of Tag. The decimal value is 1.
Unique ID of RFID Tag Set	x13 - x4	10	Unique ID of the RFID Tag. Value ranges from 1 to 1023
Absolute Loc in meters (11 1111 1111 1111 1111 : Not Applicable)	x31 - x14	18	Geographical location of the RFID Tag in terms of decameters as per Railway metrics or Location referred in the Signal Interlocking Plan of the respective stations. Value ranges from 0 to 262142
TIN in Nominal Direction	x38 - x32	7	Track Identification number encountered while traversing in Nominal direction (Incremental direction of absolute location) Value ranges from 0 to 127
TIN in Reverse Direction	x45 - x39	7	Track Identification number encountered while traversing in Reverse direction (Decremental direction of absolute location) Value ranges from 0 to 127
Station Code	x55-x46	10	For Future Use Unique ID of the Station Value ranges from 1 to 1023
Applicable Direction (0 = Nominal, 1 = Reverse)	x56	1	Denotes the applicable direction of the Signal '0' if Nominal '1' if Reverse
Signal ID in Applicable Direction	Y2-y0 & x63-x57	10	y3-y2-y1-y0-x63-x62 = 001000 : Only in RFID Tag, not in Radio Packet. Loco KAVACH should apply Brake when it crosses signal with Zero ID (dead stop locations - such as end of berthing tracks with Shunt Signals) in Normal Mode. y3-y2-y1-y0-x63-x62 = 001001 : Only in RFID Tag, not in Radio Packet. Loco KAVACH should apply Brake when it crosses signal with this 'Yard Exit' type (such as at BLOCK SECTION LIMITING BOARD (BSLB)) Other than in SR mode. Apart from above, rest codes to be used in future
Whether the signal has diverging routes	Y3 (y3 = 0: No, y3 = 1: Yes)	1	Future Use, Presently to be set '0'
ID of Approaching Signal ahead in Applicable Direction	y13-y4	10	Future Use, Presently to be set '0' Same as fields of CUR_SIG_INFO of Stn-to-Loco Radio Packet for future work
Fill Zeros	y47-y14	34	Zero padding to be done
CRC	y63-y48	16	Compute CRC (Cyclic Redundancy Checksum) to check the data integrity
Total		**128**	

2.16 Refer Annexure for Typical KAVACH Table of Control

Kavach Table of Control in annexure shows the RFID TIN-TAG layout and Control Table of Gullaguda (GGD) station of Lingampalli (LPI) - Vikarabad (VKB) section of Secunderabad division, South Central Railway.

CHAPTER

3 Communication System for KAVACH

3.1 Planning of KAVACH Towers

This chapter describes the requirements for data transmission over the air (through radio), Multiple Access scheme and Radio communication protocol for Loco and Stationary KAVACH sub-systems (Loco KAVACH and Stationary KAVACH units) and defines the Radio communication transmission time slots and frequencies required for Stationary KAVACH and Loco KAVACH system.

With the advent of high speed and high density Rail Networks over IR, implementation of Automatic Train Protection (ATP) systems has become a *sine qua non* for ensuring safety of trains. Most ATP systems provide an additional layer of safety over the existing interlocking systems so as to aid the loco pilots. All the ATP systems need radio communication towers for functioning.

3.2 Need for Communication Towers

Communication between locomotive KAVACH and stationary KAVACH is exclusively through radio waves. At present, KAVACH uses spot frequencies in 400 MHz to 450 MHz band (f_0 – 441.8 MHz, f_1 – 456.8 MHz, f_2 – 416.8 MHz, f_3 – 466.8 MHz, f_4 – 426.8 MHz). Considering the scenario where a loco is moving from block station A to block station B, it should communicate with stations A and B for approximately half the distance of block section each. Without loss of generality, a block section can safely be assumed to be around 10 km. This means that one stationary KAVACH equipment should cater to around 5 km communication. An antenna at ground level can simply not perform the job. Here arises the need for communication towers to increase the range of radio coverage.

3.3 Initial Reconnaissance Survey for Towers and their Feasibility

The most important initial step in taking up tower works is to carry out initial survey studying the feasibility of towers keeping in mind factors such as height of tower, signal strength, ease of construction and maintenance of sites. During this survey, sites could be classified into Green, Amber and Red based on the difficulty of implementation.

Classification	Green	Amber	Red
Plot Size	No limitation	Poses limitation in Tower Height/Design Selection	Serious Constraints
Plot Access	No limitation	Poses possibility of Head Load	Serious Constraints
Ground Water	No limitation	Moderate saturation posing limitation in selection of tower	Extreme saturation
Soil/Rock Conditions	Good soil and non-rocky conditions	Moderately good soil	Extreme limitation in soil strength
Summary Action	None	Tackling with Design Modifications	Find alternatives before contract is awarded.

A typical survey format detailing various geographical and technical factors for identifying tower plot location is given below.

Name of the Station / IBS / LC Gates for the proposed Tower location:

S.No.	Survey Item Description	Remarks
a	Any High Rise Buildings/trees available near proposed tower location	Yes/No
b	Type of Soil-Rocky/Clay/Black Cotton/Sand etc	
c	Soil is natural mother earth or filled soil	
d	Nearest Airport available to the station/LC gate/IB& Distance in Kms	
e	Is Station/LC gate/IB situated in Forest area	Yes/No
f	Is Station/LC gate/IB situated in Coastal area (with in 15 Kms from Coast)	Yes/No
g	Tower shall be placed 46 m (Horizontal Distance) away from the Nearest track as far as possible. Is 46 m Horizontal Railway Boundary clearance available	Available/Not Available Distance between tower and nearest track centre -
h	Availability of high tension wires in near vicinity	Yes / No
i	Any water pipelines in near vicinity of tower location	Yes / No

Contd...

S.No.	Survey Item Description	Remarks
j	Availability of staff quarters/houses nearby	Yes / No
k	Availability of underground Signalling / Telecom, / Power cables / OFC	Yes / No
l	Interdistance Distance between proposed tower and Relay room	
m	Interdistance Distance between nearest track centre to railway boundary	
o	Quad cable, OFC, signalling cable paths (same side/other side of tower location)	Quad - OFC - Signalling -
p	Longitude &Lattitude coordinates of proposed tower site for Ligowave software / WPC / MSL	Longitude - Latitude - MSL - Pincode -
q	Attachment of site photos	Yes/No
r	Motorable access to the proposed site for transportation of fabricated tower materials	Yes / No
s	Any future plans by other departments for utilisation of proposed tower plot area	
t	Any sanction of ROB / RUB in lieu of LC gate	Yes / No
u	Sanction of new interlocked LC gate	
v	Other information if any	

3.4 Selection of Tower

3.4.1 Selection of Type of Tower

Based on the initial planning, drone survey, Radio Signal Strength Indicator (RSSI), site reconnaissance survey, tower types are chosen. The types of towers are discussed below.

Self Supporting Lattice Towers – They can be angular, tubular or a hybrid of both angular and tubular. Similarly, self supporting towers are either 3 or 4 legged.

Self Supporting Monopole – It is a single self supporting pole. This design is generally used to minimise foot print at the site which is easily accessible.

Narrow Based Self Supporting Lattice Tower – It is generally a 4 legged tubular tower upto 40 m height with very low foot print. It is used as a substitute to monopole when there are no aesthetic constraints.

Characteristic	4 Legged Angular	3 Legged Angular	Narrow Based	Monopole
Foot Print	4	3	2	1
Visual Appeal	4	3	2	1
Ease of Fabrication	1	2	3	4
Ease of Erection	3	1	2	4
Overall Cost	**2**	**1**	**3**	**4**
Remarks	**1 denotes lowest / cheapest / easiest** **4 denotes highest / costliest / difficult**			

Three Legged Angular Tower

Four Legged Angular Tower

Tubular Tower

Monopole Tower

Narrow based Tower

Fig. 3.1

Monopole tower requires motorable road access to site, long boom cranes, space for horizontal assembly of monopole at site and lifting of monopole using crane. This means that monopole is not a recommended option over lattice tower at constrained site locations. At such locations, narrow based lattice towers are preferable. As such, it is opined that three legged tubular tower of 40 m height is preferable in line with RDSO guidelines to have uniformity and standardisation.

3.4.2 Selection of Height of Tower and Number of Towers

The height of tower and number of towers can be decided based on the following inputs.

(a) Geological conditions of site – Soil Bearing Capacity, Water Table, Drainage etc

(b) Accessibility of site for carrying material

(c) Radio network link calculations. The received signal strength should not be poorer than −85 dBm throughout the communication mandatory zones. It may be ensured that there are no obstructions in 60% of the First Fresnel Zone.

(d) The antennae for communication system of stationary KAVACH system shall be able to provide a minimum range of communication upto 1.5 km beyond the the first signal of the stationary KAVACH unit. This is typically 4.5 km in case of double distant territory. In the case of IBS/Mid-section Interlocked LC Gates, this is much smaller and hence, 15 m/20 m tower could also be explored based on survey details.

(e) Design Basis Report of complete route containing the following

- (i) Confirmation of tower (3 legged, angular / Tubular, Monopole, etc.,)
- (ii) Elevation of antenna on the tower structure
- (iii) Wind speed
- (iv) Terrain condition
- (v) Seismic conditions
- (vi) IR Code of Practice issued by RDSO to be followed for structural designs, fabrication and erection of towers. Reference to RDSO's B&S Directorate checklist on Design of Super-structure for KAVACH tower vide document dated RDSO-BnS0EBS(SB-2)/16/2020-O/o ED/BnS/RDSO dated 11/01/2021.
- (vii) Material specification

(f) Above details can be finalized by

- (i) Reconnaissance (Walk over survey)
- (ii) Ligowave software to decide the antenna height
- (iii) Drone survey to have the exact locations of stations, Signal structure, IBS and Auto goomties, kilometer stones, traction masts etc.

(g) RSSI (Received Signal Strength Indicator) Survey is required to fine tune the antennas light, which is already available with the help of Ligowave software using the longitude and latitude of tower location at stations, IBS and Auto-goomties, LC gates, etc.

(h) If we fine tune the antennas height also in the beginning, then only RSSI survey in advance of inviting EPC tender is required.

Site Information			
TX Site Name	Secundrabad Jn	RX Site Name	Sitafalmandi
Radio Type	LigoPTP 6-N RapidFire	Radio Type	LigoPTP 5-23 RapidFire
Latitude	17.434	Latitude	17.428
Longitude	78.508	Longitude	78.520
TX Power	40.0 dBm	RX Threshold	-85.0 dBm
Ant. Gain	5.0 dBi	Ant. Gain	3.0 dBi
Ant. Height	35.0 meter	Ant. Height	5.0 meter
Parameters			
Frequency	441.0 MHz	Climate	Continental Temperate
Ant. Polarization	Vertical	Measurement System	Metric System
Misc. Loss	5.0 dBm	Rain Rate	0.0 mm/hr
Results			
Total Path Loss	93 dB	Thermal Fade Margin	40 dB
RX Signal Level	-44.879 dBm	Distance between sites	1.437 km
EIRP	45.0 dBm	Link availability due to rain	N/A

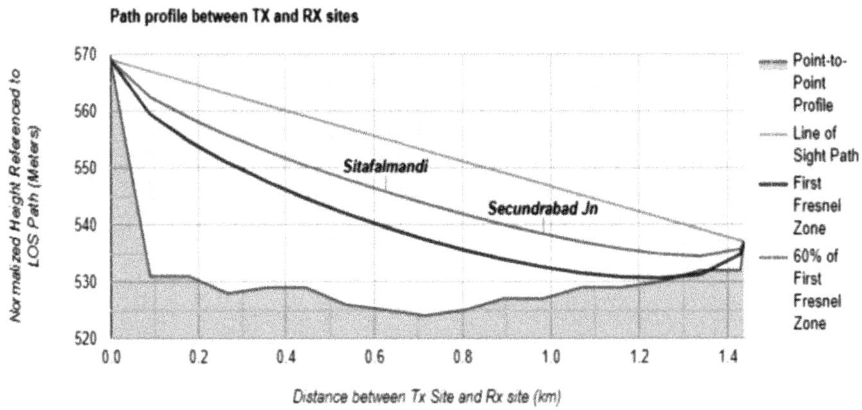

Fig. 3.2 A Sample RSSI Survey using Ligowave Software

3.5 Design Engineering Activities

3.5.1 Design Approval Process

Design approval process can be summarised as below:

(a) Preparation of designs of tower sub-structure (foundation) and super-structure (tower) based on Wind Speed (IS 875 Part 3), Soil Bearing Capacity (SBC), Water Table (With/Without Saturation), Factor of Safety (2.0 as per IR Standard

Code of Practice for the Structural Design of Microwave Towers of Self Supporting Type, 1982) etc.

(b) Proof checking of design by third party like CPRI/IITs.

(c) Railway approval of design by Chief Bridge Engineer (CBE).

(d) The tower designs are specific to each design agency and hence, separate approvals need to be obtained for each agency.

3.5.2 Site Specific Activities

(a) Identification of site for tower erection free from infringements and approval by division duly recommended by all stakeholders/departments.

(b) Collection of soil samples at the identified locations and their lab testing

(c) Submission of Soil Investigation Report stating SBC to railways along with relevant foundation drawing for review and approval.

(d) Final approval for "Fit for Construction" by project executing unit.

3.5.3 Site Execution Activities and Timelines

Team	Activity	Team Skill Set	Duration	Remarks
A	Site readiness and Mobilization towards work commen-cement	Project management and Site coordination	10 days	Involves interaction with multiple stake holders such as verification of UG cable and overhead lines, Vegetation clearance, Site leveling and marking with Railway clearances.
B	Civil foundation construction	Skilled workers for concreting, steel bar bending, shuttering activities	10 days	Activities include: Site excavation ~ 8 × 8 × 3 meters PCC bed concreting Placing of reinforcement steel. Concreting of raft and first lift column Concreting of second lift column Concreting of third and final lift with CIP Stage-wise inspection at critical milestone is carried out and recorded in site register by Railways
C	Tower erection	Skilled tower rigger	8 days	Erection of tower including ladder and cable tray working and equipment platform, staging, antenna mounts, lightning arrester/ SPDs, aviation lamp, GI earth strip along the tower, etc.
D	Tower painting	Skilled tower rigger with painting skills	4 days	Activity includes Primer coat using zinc chromate. Two coats of enamel paint alternate deep orange and white paints.

Contd...

Team	Activity	Team Skill Set	Duration	Remarks
E	Manual Curing using open pit and watering during various stages of foundation construction	Unskilled / Semi skilled labour and Supervisor	12 days	This activity timelines can be compressed subject to implementation of certified curing compounds during concreting.
F	Tower completion and verticality check	Skilled team along with RITES	1 day	Tower ready for installation of antennas, cabling, and Radio modems, levelling, fencing, ring earthing, etc.,

Note: Tower construction would take approximately 45 days per tower. By engaging four teams each of Civil, Tower and Site supervision, approximately 12 towers sites can be completed in a month with concurrent management. However, the above parallel activities will become effective after initial period of 6 to 8 weeks from the commencement of the project activities.

3.6 Inspections and Testing of Tower

3.6.1 Inspection and Tests for Foundation and Tower

S.No.	Description	Items for Testing	Remarks
1	Tests on Foundation		
1.1	Raw Material	Reinforcement sample for Tensile strength as per IS 1786	Laboratory report to be submitted
		Coarse and Fine Aggregate for fineness modulus as per IS 383	
		Cement conforming to IS standard for strength as per IS 269	
1.2	During foundation works	Slump Cone test on fresh concrete for workability to be performed at site during each stage of concreting as per IS 456	Slump value shall be between 75 to 90 and witness by Railway
		Collection of Cube samples from fresh concrete for strength check to be performed at site during each stage of concreting as per IS 456	1 sample (3 specimen) for every 5 Cu.m to be tested at lab after 28 days and report to be submitted to Railways

Contd...

S.No.	Description	Items for Testing	Remarks
2	Inspections on Foundation		
2.1	During foundation works	1st inspection – Depth check of foundation and layout of foundation at PCC	Witness by Railway and recorded in Site Register
		2nd inspection – Reinforcement check before casting of Footing and supervision of concreting of footing and 1st lift of column / pedestal	
		3rd inspection – Checking of Template fixing and supervision of concreting of final lift of column / pedestal	
3	Tests on Tower – at Factory by RITES as per Quality assurance Plan (QAP)		
3.1	Raw material to workmanship	The Tests at the Factory by RITES shall be as per agreed QAP – Refer table below for details	RITES inspector shall visit the Factory for Raw material inspection and Finished Goods inspection and submit the certificate of clearance.
4	Inspections on Tower – at Site		
4.1	Material inspection at site	4th inspection - Checking of materials for good condition, quantity as per packing list and coating thickness	Railway inspector to verify and record observations in site register
4.2	Tower Post erection	5th inspection – Checking of tightening of bolts (Randomly) and verticality check.	

3.6.2 Tower Super-structure Material Inspection by RITES – Quality Assurance Plan (QAP)

Tower super-structure material inspection is done by RITES at two stages.

1. Raw material inspection for super-structure of tower
2. Inspection of finished goods (members) of tower super-structure

RITES inspection shall be carried out in accordance with Quality Assurance Plan (QAP) that has been prepared and agreed by all the stakeholders – Railway, Tower Designer, Contractor Agency and Tower Fabricator. Necessary approvals shall be obtained from CBE. The broad guidelines of QAP and manufacturing process right from raw material to finished product are elaborated below.

S. No.	Activity	Description	Reference IS Codes - For Supply and Acceptance Criteria	Remarks
1	Raw Material Procurement	Pipes, Plates, Angles, ISMC, Hardware (HDG Bolts of Gr 8.8 and nuts)	IS 2062 for all hot rolled sections; IS 1161 / IS 1239 / IS 3601 for Pipes IS 12427 Or other equivalent	Along with supply MTC is normally expected; Also samples are usually collected and tested for Chemical and Mechanical properties at the time fabrication as a usual process

Contd...

S. No.	Activity	Description	Reference IS Codes - For Supply and Acceptance Criteria	Remarks
2	In Process Fabrication	Dimensional cut of members, assemblies, hole punching / drilling	IS 10748; IS 2062; IS 1161; IS 7215	Use of Jigs and Fixtures for assembly of sections shall be ensured.
3	Welding	Welding shall be carried out in accordance to WPS / Drawings	IS 816; IS 822; AWS D1.1	Welding shall be visually inspected and samples randomly tested by NDT methods of DPT / MPI
4	Final Inspection	Inspection of fabricated tower	Drawings	Post Fabrication final inspection of material is usually carried out
5	Galvanization	Galvanization of tower	IS 6745; IS 2633; IS 2629; Drawings	Coating thickness shall be as per drawing usually between 45 to 85 Microns depending on thickness
6	Packing and Loading	Tower despatch to site	Packing List and Drawings	Loading shall be inspected and supply quantities shall be ensured as per drawing

3.6.3 Site Photographs of Ideal Conditions

Fig. 3.3 Typical Tower Site Photographs

3.7 Cable Connectivity for Tower at Stationary KAVACH Locations

Cable shall be provided at stationary KAVACH locations for tower catering to the power supply radio modems, aviation lamp and RF transmission of data. Cable connectivity diagram is shown below.

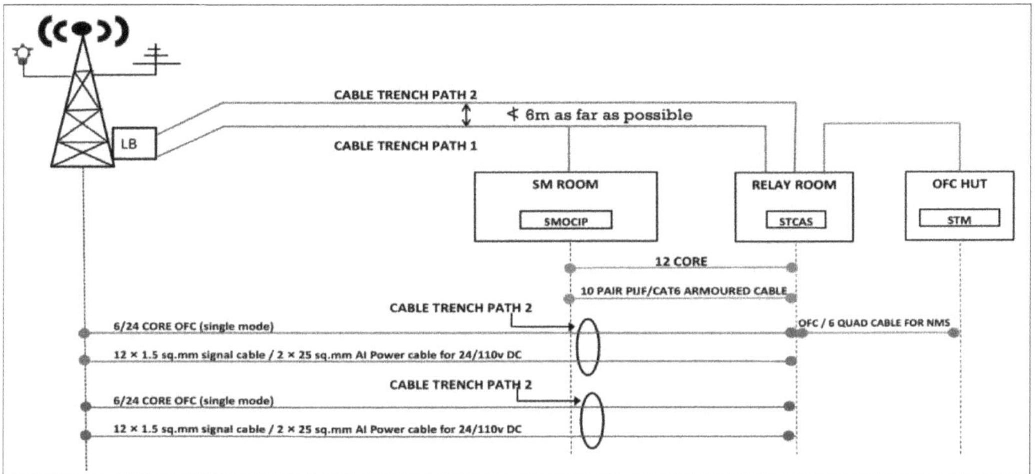

Fig. 3.4 Cable Connectivity Diagram for KAVACH at Stationary KAVACH Locations

Notes for Figure 8

1. Two diversified cable paths shall be provided as far as possible, from Station KAVACH to the tower to avoid common mode failures.
2. 6/24 OFC cable in diverse path shall be provided from Stationary KAVACH to the tower to avoid common mode failures.
3. 12 Core Signalling cable or Power cable for operating Radio of aviation lamp.
4. 12 Core Signalling cable between SMOCIP to Stationary KAVACH.
5. OFC cable for NMS purpose
6. LMR-600 RF coaxial cable from the two radio modems to antennae shall be routed in the different paths.

3.8 WPC / SACFA Clearance Process

Wireless Planning and Coordination (WPC) clearance is necessary for the operation of radios in India. This clearance is for the operation of radio. Site specific clearances such as defence areas, airport area etc., are given through SACFA clearance. Revised process for the application is given below.

(a) SARAL SANCHAR (Simplified Application For Registration and Licenses) a Web based Portal, for Issuing of various types of Licenses and Registration certificates is part of various Digital initiatives being taken by Department of Telecommunications. (http://saralsanchar.gov.in/)

(b) Applications to be processed in SARAL SANCHAR online portal through sub-user creation
(c) A Master User has been created for Ministry of Railways for creation of Sub-Users on Saral Platform of WPC (SACFA), who can then apply further for allotment of frequencies
(d) KAVACH uses FIVE frequency spots in 400 MHz to 450 MHz band. (f0 - 441.8 MHz, f1 - 456.8 MHz, f2 - 416.8 MHz, f3 - 466.8 MHz and f4 - 426.8MHz).
(e) SCR has initially processed applications for the WPC clearance for KAVACH frequencies during December 2020 after introduction of Portal.
(f) Two Radios for each Stationary KAVACH/LC KAVACH and Two radios for each Loco KAVACH and 10% Spare radios
(g) There is a restriction of maximum 99 nos. of radios per application in the portal, hence processed with multiple applications (Six) to cater the entire 1200km on-going KAVACH project.
(h) Approval process takes 6 to 8 months and hence Railways shall initiate the process well in advance
(i) With Dealer Possession License (DPL), OEMs can import KAVACH Radios, but for supply and operation WPC license is required.
(j) Main aim of Standing Advisory Committee on Radio Frequency Allocation (SACFA) is to ensure aviation safety and security and addressed by clearance for a tower/antenna/site (Heights> 30 Meters)
(k) Initiation of SACFA clearance after obtaining the Decision letter from WPC which is a pre- requisite to fill the online application for SACFA clearance in the portal.

3.9 Maintenance of Towers

Once KAVACH towers are erected, it is essential that they are kept in "fit" condition through regular maintenance. It generally consists of general inspection, fault tracing, checking of tower aviation lights, quarterly checking of condition of foundation, corrosion, tilt, nut and bolt tightness, distortion of members, cabling intactness, earthing, half yearly verticality check and other need based maintenance. Painting of tower is also needed every three years.

3.10 Conclusion

Successful deployment of KAVACH needs timely erection of communication tower for establishing connectivity between Stationary KAVACH and Loco KAVACH. One of the critical activities of KAVACH deployment is construction of towers and none of the testing activities could be initiated without the erection of towers. Initial reconnaissance survey for towers and their feasibility, meticulous planning, regular progress reviews with tower contractors, resolving site issues, concurrent management of tower activities are some of the key areas for meeting the targets. ADEN or

SSE/Works possessing adequate competency in Civil foundations and steel structures shall be deputed under the administrative control of S&T department for authorised working.

Station Antenna

Base station stacked Dipole type Omni directional type antenna with Gain of 9.0 dB. Frequency band is 406-512 MHZ with input impedance 50 Ohms. Four antennae used and 1- TX and 1- RX antenna for each modem.

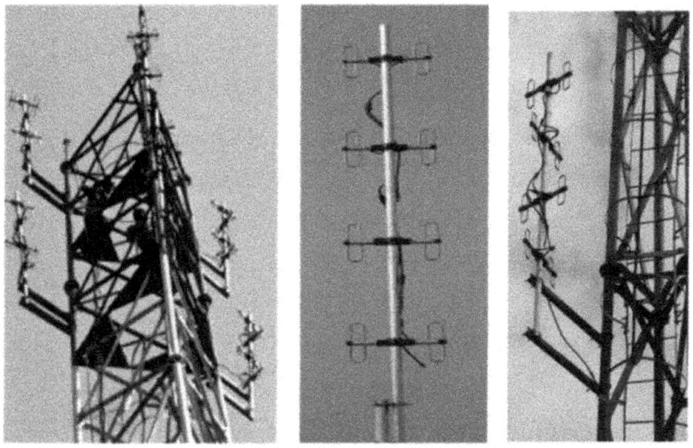

Fig. 3.5 Antenna Towers

LOCO – Radio Antenna:

Fig. 3.6 LOCO – Radio Antenna

1. Gain 0 dB, Frequency band : 406 – 512 Mhz
2. Input Impedance: 50 Ohm
3. Polarization: Vertical
4. Radiation pattern: Omni directional
5. Wind rating: 200 km/hr
6. Weight (Kg): 3
7. 4 Antennae (1-TX and 1-RX antenna for each radio modem)

3.11 Fiber Classification, Standards & Constructional Features

3.11.1 Fiber Classification

Fiber Classification based on fiber types in a cable may be categorized as three types. They are:

1. Single Mode Fiber Optic Cable.
2. Multimode Fiber Optic Cable.
3. Hybrid/Composite Cable.

1. **Single Mode Fiber Optic Cable:** Designs of single-mode fiber have evolved over several decades. The three principle types and their ITU-T specifications are:

- Non dispersion-shifted fiber (NDSF) G.652

  The initially deployed type used for 1310 nm. This fiber has high dispersion at 1550 nm, hence not suitable for 1550 nm systems. To solve the shortcoming of NDSF fiber, fiber manufacturers developed dispersion-shifted fiber (DSF).

- Dispersion-shifted fiber (DSF) G.653

  In this type of fiber the fiber is designed in such a way that the zero-dispersion point is moved to the 1550 nm region

- Non zero-dispersion-shifted fibers (NZ-DSF) G.655

  Though DSF worked extremely well with a single 1550 nm wavelength, it exhibits serious nonlinearities when multiple, closely-spaced wavelengths in the 1550 nm were transmitted in DWDM systems.

  To address the problem of nonlinearities, non-zero-dispersion-shifted fibers (NZ-DSF) were designed by manufacturers. The fiber is available in both positive and negative dispersion varieties and is rapidly becoming the fiber of choice in new fiber deployment.

2. **Multimode Fiber Optic Cable:** All fibers in the cable are multimode cables fibers only. Multi-mode optical fiber is a type of optical fiber mostly used for communication over short distances, such as within a building or on a campus. Multi-mode links can be used for data rates up to 100 Gbit/s. Multi-mode fiber has a fairly large core diameter (50/62.5μm) that enables multiple light modes to be propagated and limits the maximum length of a transmission link because of modal dispersion. The standard G.651.1 defines the most widely used forms of multi-mode optical fiber.

3. **Hybrid/Composite Cable:** Both single mode and multimode fibers are packaged in one cable, such as 4 multimode fibers and 4 single mode fibers in a single cable.

   Classification based upon application may be categorized as the following types. They are:

   1. Indoor Cables
   2. Outdoor fiber optic Cables
   3. Direct buried Outdoor Cables

1. ***Indoor Cables*:** A single cable structure with a single fiber. Simplex cable varieties include 1.6mm & 3mm jacket sizes. Fire safety is the number one factor in selecting indoor cables, particularly those that run through plenum spaces. Indoor cables must pass the flame-retardant and smoke-inhibitor ratings.

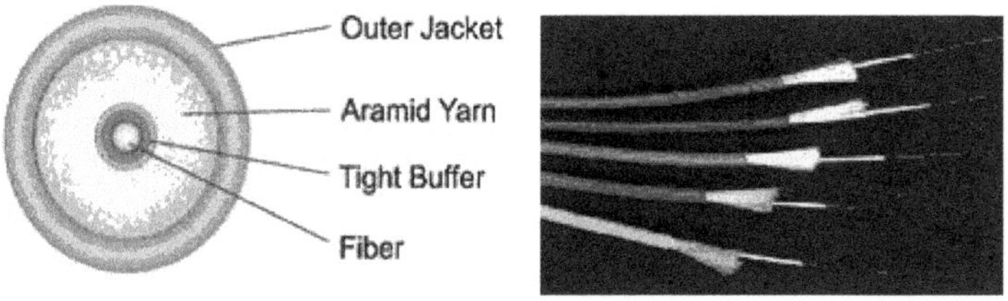

Fig. 3.7 A Indoor OFC

2. ***Outdoor Fiber Cable*:** Outdoor rated tight buffered cables have riser and plenum rated versions. These cables are flexible, easy to handle and simple to install. Since they do not use gel, the connectors can be terminated directly onto the 900um fiber without difficult-to-use kits. This provides an easy and overall less expensive installation.

Fig. 3.7 B Outdoor Tight Buffered Fiber Optic cable

3. ***Direct-buried (Armored Fiber Optic Cable):*** Armored cables are similar to outdoor cables but include an outer armor layer for mechanical protection and to prevent damage. They can be installed in ducts or aerially or directly buried underground. Armor is surrounded by a polyethylene jacket.

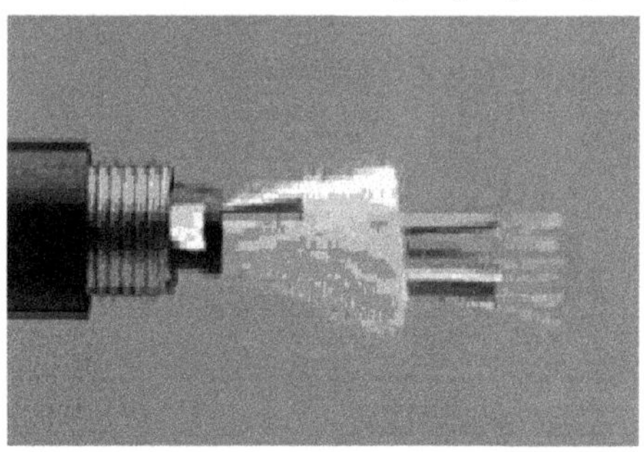

Fig. 3.7 C Armored Fiber Optic cable

Armored cable can be used for rodent protection in direct burial if required. This cable is non-gel filled and can also be used in aerial applications. The armor can be removed leaving the inner cable suitable for any indoor/outdoor use. Temperature rating −40 to +85^{0}C.

3.11.2 Standards and commercial fibers

The physical dimensions of core and cladding have been standardized to ensure compatibility among splices and connectors.

The international standards for SM fiber are:

Cladding diameter : 125 microns (micro meter)

Cladding + coating : 245 microns (micro meter)

Core diameter : 8 to 10 micro meter

The International standards for MM fiber are:

Cladding diameter : 125 microns (micro meter)

Cladding + coating : 245 microns (micro meter)

Core diameter : 50 to 62.5 micro meters

The SM and MM Core and cladding diameters:

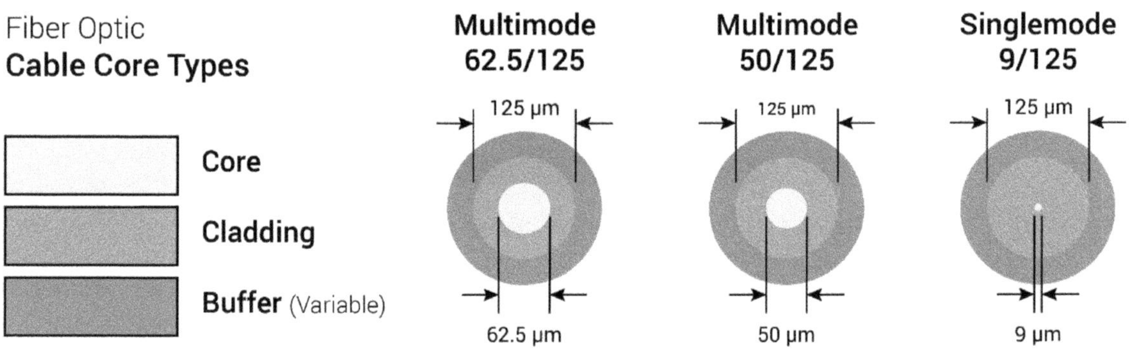

Fig. 3.8 Core and Cladding diameter

Technical Specifications: Technical Specifications are as per ITU-TG.652.D and tabulated below:

Cut off wave length	Cut off wave length for 2m fiber section shall be 1320nm
Attenuationat 1310 nm between 1285 to 1330 nm at 1550 nm	≤ 0.36 dB/Km < 0.36+0.02dB/Km ≤ 0.23 dB/Km
Nominal diameter of core	8.8 μm - 9.8 μm for matched clad fiber
Nominal diameter of cladding	125 μm ± 1.0 μm
Non circularity of cladding	≤1%
Mode field concentricity error	≤0.8 μm

Contd...

Cut off wave length	Cut off wave length for 2m fiber section shall be 1320nm
Primary coating	Material UV curable Acrylate, Diameter 245±10 μm. It should not have any reaction with cladding or cable material
Chromatic dispersion(CD)	@1550nm: ≤18.0 , @1625nm: ≤22.0 (ps/(nm.km))
Polarization mode dispersion (PMD)	For individual fiber : ≤0.1(ps/(km)$^{1/2}$ For cabled fiber : ≤0.4(ps/(km)$^{1/2}$
Fiber curl	≥ 4m radius of curvature
Fiber micro bend at 1550 nm	≤ 0.5 dB

3.11.3 Constructional Features

Marking on Cable

On the outer jacket of the cable there is indelible length marking at an interval not exceeding 1±.01 meter. The outer jacket is in black color and marking is in white color.

The sequential length marking shall not rub off during normal installation. The marking on the cable is as given below:

"Manufacturer/Company's trade mark, IR marking, telephone mark, laser symbol, type of cable, number of fibers, month, year of manufacture and drum number."

Identification of Fibers in Cable

Fibers are color coded with readily distinguishable& durable colors. In case of four fibers in a tube the order of colored fibers are Blue, Orange, Green and Natural. The 6 loose tubes have the following colors.

Loose Tube number	Color of loose tube
1	Blue
2	Orange
3	Green
4	Brown
5	Slate
6	White

3.12 Multiple access Techniques in Communication

FDMA

FDMA is the process of dividing one channel or bandwidth into multiple individual bands, each for use by a single user. Each individual band or channel is wide enough to accommodate the signal spectra of the transmissions to be propagated. The data to be transmitted is modulated on to each subcarrier, and all of them are linearly mixed together.

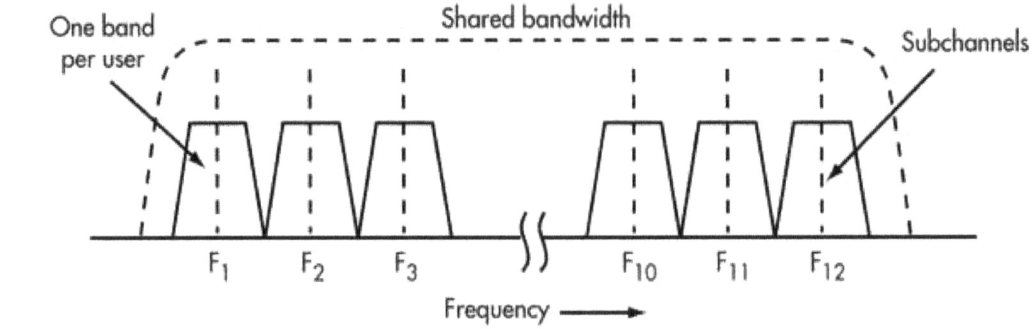

Fig. 3.9

TDMA

TDMA is a digital technique that divides a single channel or band into time slots. Each time slot is used to transmit one byte or another digital segment of each signal in sequential serial data format. This technique works well with slow voice data signals, but it's also useful for compressed video and other high-speed data.

A good example is the widely used T1 transmission system, which has been used for years in the telecom industry. T1 lines carry up to 24 individual voice telephone calls on a single line. Each voice signal usually covers 300 Hz to 3000 Hz and is digitized at an 8-kHz rate, which is just a bit more than the minimal Nyquist rate of two times the highest-frequency component needed to retain all the analog content.

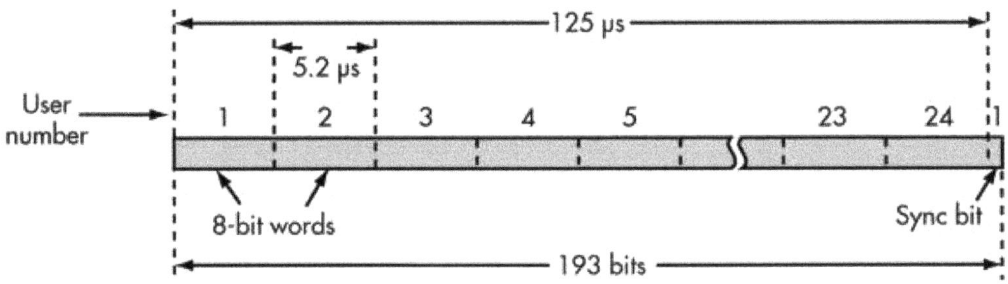

Fig. 3.10

CDMA

CDMA is another pure digital technique. It is also known as spread spectrum because it takes the digitized version of an analog signal and spreads it out over a wider bandwidth at a lower power level. This method is called direct sequence spread spectrum (DSSS) as well. The digitized and compressed voice signal in serial data form is spread by processing it in an XOR circuit along with a chipping signal at a much higher frequency. In the cdma IS-95 standard, a 1.2288-Mbit/s chipping signal spreads the digitized compressed voice at 13 kbits/s.

The chipping signal is derived from a pseudorandom code generator that assigns a unique code to each user of the channel. This code spreads the voice signal over a bandwidth of 1.25 MHz. The resulting signal is at a low power level and appears more like noise. Many such signals can occupy the same channel simultaneously. For example, using 64 unique chipping codes allows up to 64 users to occupy the same 1.25-MHz channel at the same time. At the receiver, a correlating circuit finds and identifies a specific caller's code and recovers it.

The third generation (3G) cell-phone technology called wideband CDMA (WCDMA) uses a similar method with compressed voice and 3.84-Mbit/s chipping codes in a 5-MHz channel to allow multiple users to share the same band.

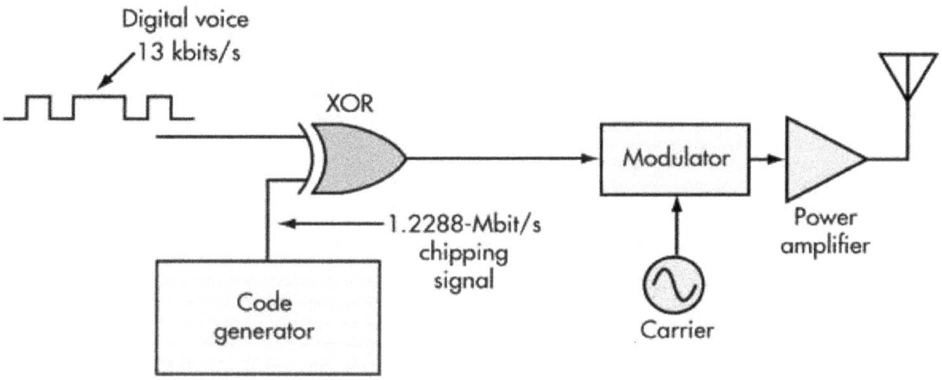

Fig. 3.11

The chipping signal is derived from a pseudorandom code generator that assigns a unique code to each user of the channel. This code spreads the voice signal over a bandwidth of 1.25 MHz. The resulting signal is at a low power level and appears more like noise. Many such signals can occupy the same channel simultaneously. For example, using 64 unique chipping codes allows up to 64 users to occupy the same 1.25-MHz channel at the same time. At the receiver, a correlating circuit finds and identifies a specific caller's code and recovers it.

The third generation (3G) cell-phone technology called wideband CDMA (WCDMA) uses a similar method with compressed voice and 3.84-Mbit/s chipping codes in a 5-MHz channel to allow multiple users to share the same band.

OFDMA

OFDMA is the access technique used in Long-Term Evolution (LTE) cellular systems to accommodate multiple users in a given bandwidth. Orthogonal frequency division multiplexing (OFDM) is a modulation method that divides a channel into multiple narrow orthogonal bands that are spaced so they don't interfere with one another. Each band is divided into hundreds or even thousands of 15-kHz wide subcarriers.

The data to be transmitted is divided into many lower-speed bit streams and modulated onto the subcarriers. Time slots within each subchannel data stream are used to package the data to be transmitted. This technique is very spectrally efficient, so it provides very high data rates. It also is less affected by multipath propagation effects.

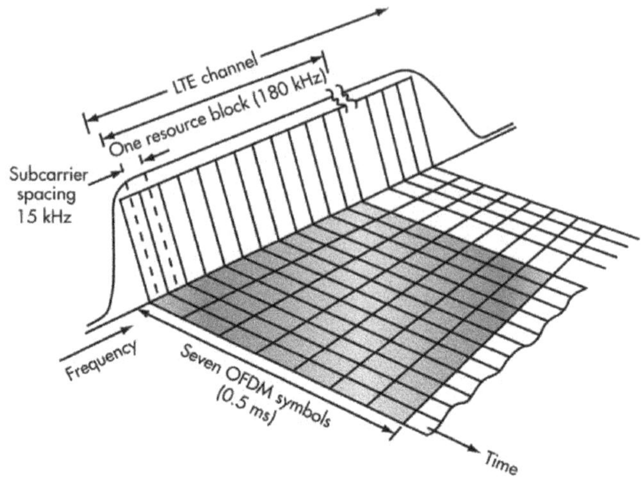

Fig. 3.12

3.13 Over the Air (OTA) Requirements

- Communication between the Stationary KAVACH and Loco KAVACH shall be Over-The- Air (OTA) using Multiple Access
- Each Multiple Access frame cycle shall be of 2000 milli seconds.
- It shall be suitable for communication in frequency range of 406 MHz to 470 MHz.
- Loco KAVACH shall use transmit frequency (f_0) in block section and at the times of emergency situations (SoS, head-on, rear-end collisions).
- Stationary KAVACH and Loco KAVACH shall use their respective timeslot(s) in the Multiple Access within their channel for the transmission of communication packet(s).
- Communication packets shall be of a maximum size of 1024 bits.
- KAVACH sub-system shall split the packet into multiple packets if it exceeds maximum communication packet size (1024 bits).
- The transmission Over-The-Air from Radio shall be controlled by KAVACH Sub-systems using Request To Send signal in the RS232.
- KAVACH sub-system may transmit multiple Communication packets in a single transmission burst.
- KAVACH sub-system shall transfer all the data for a single transmission burst to the Radio modem at least 20milli second before commencement of Over-The-Air (OTA).

- KAVACH sub-system shall calculate the timings from commencement and completion of transmission by Radio Modem Over-The-Air (OTA) by considering the preamble, communication packet, post amble and extra stuffed bits.
- KAVACH sub-system shall disable the RTS signal after completion of transmission of information data over-The-Air (OTA).
- Change of frequency or switch between the frequencies shall be completed well in Advance (15 milliseconds before commencement of Data transmission Over-The-Air (OTA)).
- In the bit-stream Over-The-Air, LSB shall be transmitted first.

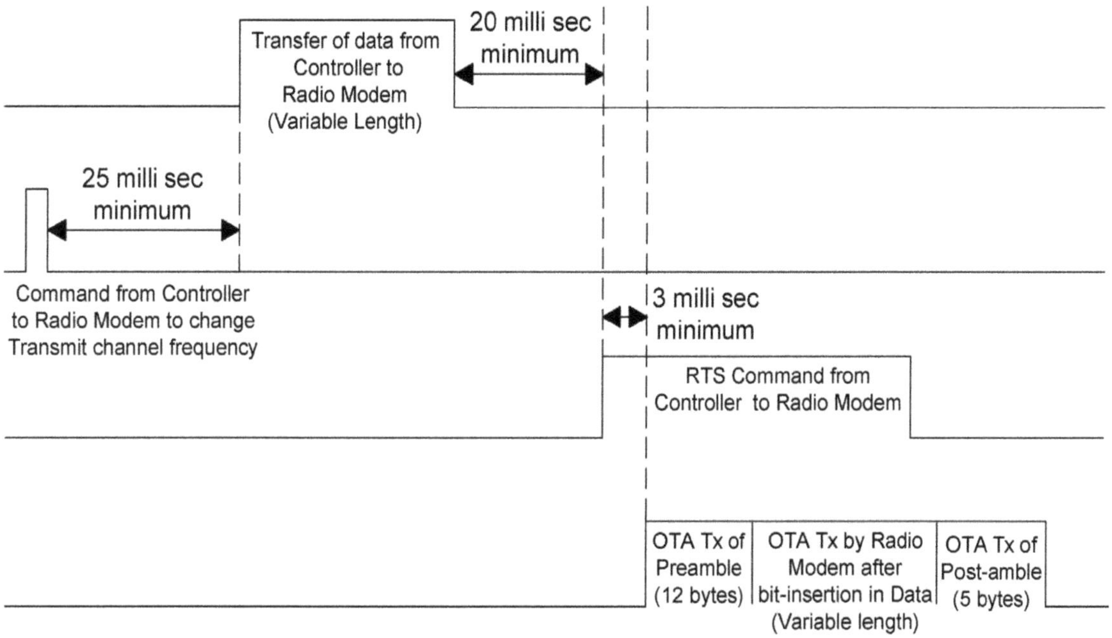

Fig. 3.13 Radio Communication

3.14 Multiple Access Scheme

The optimized frame cycle structure for the TDMA/FDMA/SDMA scheme is shown in figure 4.2, as shown frame cycle is divided into basic 78 time slot position markers (position nos. 1 to 78) each of width 352 bits (18.33 m-sec). These are spaced 96 bits (5 m-sec) apart except for the four wider time slots to ensure proper frequency stabilization on change.

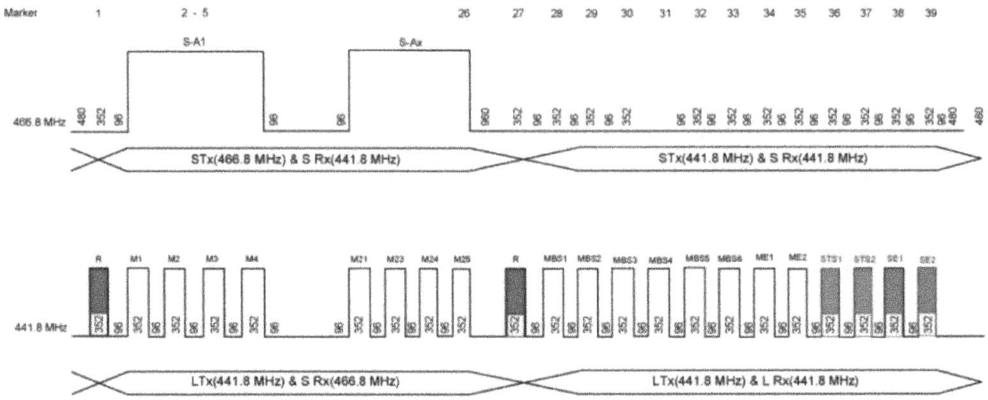

Fig. 3.14

Sl. No.	Time slots	Used for	Marked as	Remarks
1	P2 to P27	Stationary KAVACH and Loco KAVACH communication	M-1 to M-50	Total 50 Timeslots out of 78
2	P28 to P33 and P67 to P72	Loco KAVACH for broad casting communication packets in the block section	MBS-1 to MBS-12	Total 12 Timeslots out of 78. Shall switch its Tx frequency to f_0 in the block section
3	P34,35,73,74	In f_0 shall be used by the Loco KAVACH for broad casting additional emergency (SoS) messages.	ME-1 to ME-4.	Total 4 Timeslots out of 78.
4	P38,39,77,78	In f_0 shall be used by the Stationary KAVACH for broad casting additional emergency (SoS) messages	SE-1 to SE-4.	Total 4 Timeslots out of 78.
5	P36,37,75,76	In f_0 shall be used by the Stationary KAVACH for broadcasting Access Authority messages.	STS-1 to STS-4.	Total 4 Timeslots out of 78.
6	Four time Slots P1,P27,P40 & P66	Used for frequency stabilization and frequency switching	------	Total 4 Timeslots out of 78.

* Stationary KAVACH System shall transmit the Radio packet in its designated time slot*Loco KAVACH System shall transmit the radio packet in its designated time slot and designated frequency channel received from stationary KAVACH Unit. * The frequency pair allotted for two adjacent stations shall be different. * The Loco time slot for the same station shall not be adjacent to each other (minimum one time slot gap shall be kept).

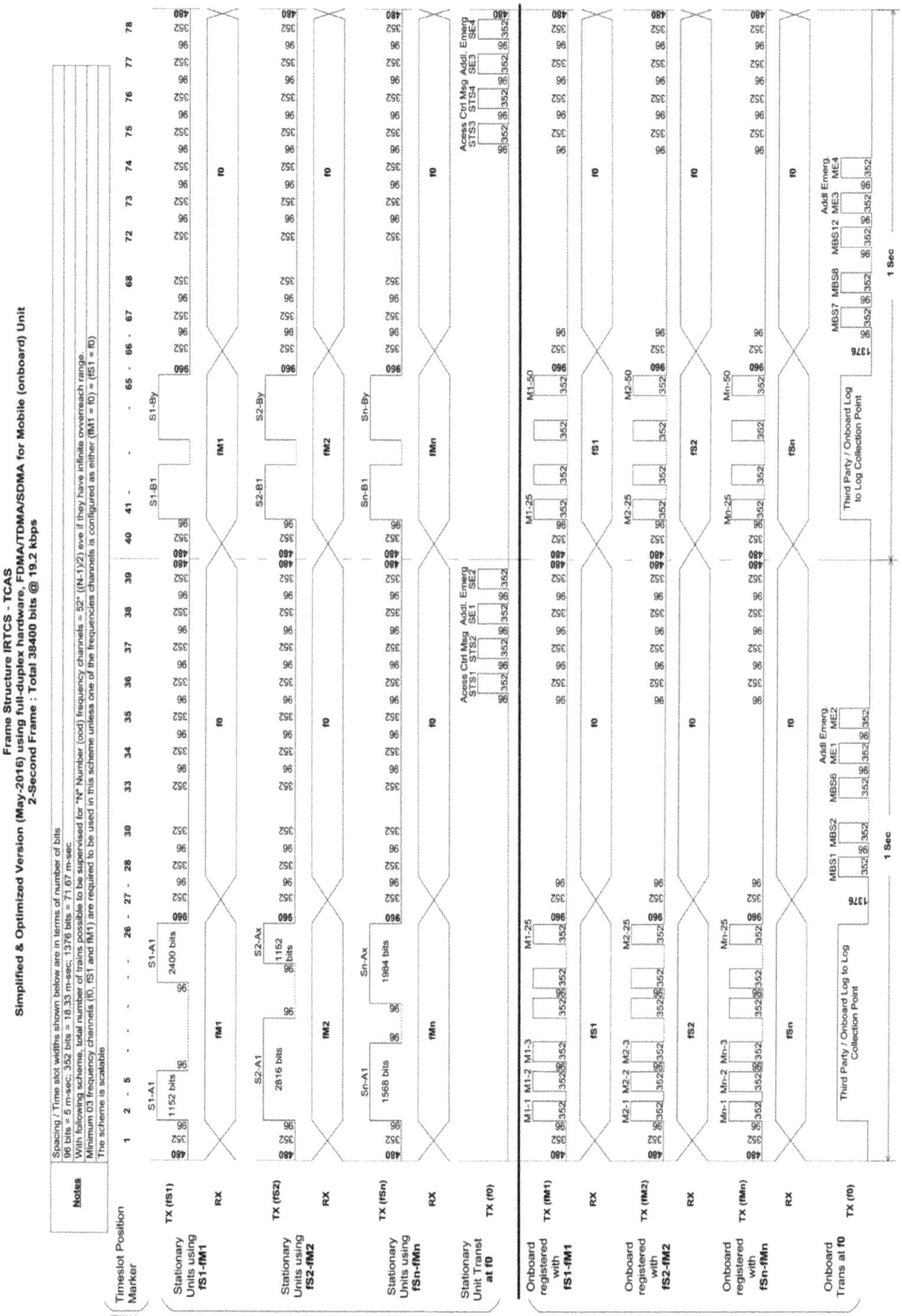

Fig. 3.15 Multiple Access Scheme

3.15 Radio Modem Requirements

Radio Unit shall be as per following specifications:

(a) Shall be FCC or IC or CE certified

(b) Shall possess RTS/CTS and/or DOX modes

(c) Shall be capable of RF Data Transfer in "Bitwise"/streaming Mode

(d) RF frequency range:406-470MHz

(e) RF Channel Bandwidth : 25kHz

(f) Modes of operation Full Duplex

(g) Modulation: 2FSK at 19200 bps with linear 8th order low pass filter (raised – cosine alpha.1Approximation).

(h) Deviation: 4.3 kHz +/- 0.1 kHz. Occupied Bandwidth : 16.35 kHz +/- 0.15kHz.

(i) Operating frequencies : Ranging from 406 MHz to 470MHz

(j) Emission : according to16K0F2D

(k) Transmitter freq. stability : 1ppm

(l) Transmitter Turn-on time (Tx. Freq. stable)/Channel Switching time: not more than 15msec

(m) Carrier Output Power: 1-10 w adjustable through software.

(n) Receiver Adjacent Channel Rejection 70dB at 25kHz

3.16 Receiver Sensitivity: 35 micro-volts for 12 dB SINAD

(a) 1 x 10^{-6} BER at -100 dBm Level for 19.2kbps and 25kHz Bandwidth

(b) Interfaces: RS232

(c) RF Impedance : 50ohm

(d) Power Supply : 10V-30V DC

(e) Set-up and Diagnostic features to be available through separate port RS232 and real time non-intrusive online diagnostics.

3.17 Functional Requirements

3.17.1 Modulation

(i) The modulation used shall be 2FSK with 19,200 baud rate with linear 8th order low pass filter (raised cosine alpha 1approximation).

(ii) Occupied bandwidth shall be 16.35 kHz +/.15 kHz

(iii) The nominal deviation shall be 4.3 kHz +/-0.1 kHz.

3.17.2 Transmission

(i) During bit stream over the air transmission, LSB shall be transmitted first.

(ii) Transmission shall start within 3ms +/- 1ms after data terminal equipment causes the signal on RTS line to be high.

(iii) RTS shall be raised before commencement of preamble transmission.

Radio modem shall transmit based on the DTR, RTS and RI signals as below:

DTS	RTS	Ring Indicator	Radio Modem
Low	*	Low	Won't transmit
High	Low	Low	Receiving or buffering Tx data
High	↑	↑	Transmit all buffered data and incoming data
High	High	High	Send all data in TX buffer and continue transmitting even when TX buffer is empty
High	↓	↓	Continue transmitting remaining data in TX buffer, then unkey
↓	High	↓	Abort transmission, discard data in TX buffer and unkey immediately

* : Don't Care; ↑ : Transition from low to high; ↓ : Transition from high to low

3.17.3 Encoding

- The radio modem shall commence transmission by prefixing preamble (12 bytes of 0x7E) to the data received from DTE.
- The radio modem shall complete transmission by suffixing post amble (5 bytes of 0x7E) to the modified data.
- The data to be sent shall be encoded as shown in following Pseudo code. Encoder state shall be updated throughout the transmission.
    - Input _Bit = Input_bit XOR 1
    - Encoder State = Encoder State XOR Input_Bit
    - Output_Bit = Encoder State

Examples:

Case	Consecutive Flag Characters	Two bytes of User Data (having all '0's)	Two bytes of User Data having all '1's with '0' stuffing
Input bit stream	0111111001111110	0000000000000000	1111101111101111101
Output bit stream	01111111011111110	01010101010101010	00000011111100000011

Radio modem shall insert additional '0' after five consecutive '1's of data during transmission.

For example,

0x7C – 01111100 is sent OTA as011111000

0xF8 – 11111000 is sent OTA as111110000

0x7E – 01111110 is sent OTA as011111010

0xFE – 11111110 is sent OTA as111110110

The encoded data shall be scrambled before stuffing of '0' bit as shown in following Pseudo code. Scrambler state shall be updated throughout the transmission.

3.17.4 Receiving

- The received data shall contain application data as well as preamble and post amble.
- Reception of complete post amble (5 bytes of 0x7E) shall act as delimiter between two successive "Receive" bursts.
- At the end of transfer of received data from radio modem to DTE, the radio modem shall additionally append "0xA5 – 0xC9 – 0xA5 – 0xC9" after the data
- After the data transfer to DTE, the EIA 232F function shall be switched to high from low, shall remain high for 2 ms and shall be switched to low again.
- Refer the below timing diagram, for data transfer between KAVACH sub- system and radio modem.

3.18 LTE Radio Modem Requirements

Introduction: The LTE Radio modem requirements to be used for the purpose of future evaluation of KAVACH System.

- LTE modems Dual SIM operation on one modem card will not provide sufficient level of redundancy.
- QoS on Uplink: UL QCI obeys UL TFT using Network Initiated QoS as per 3GPP TS24.008
- Operating system – Shall allow the use of a hypervisor to allows multiple operating systems to share the hardware host.
- Voltage – 24VDC or 110VDC.
- Temperature Operating Range :-40^0C to $+70^0$C ($+85^0$C for 10 mins) as per
  - EN 50155 Tx Compliance Heat, Cold, Insulation, Shock and Vibration:
  - EN50155:2017
  - Railway Applications - Electronic equipment used on rolling stock
  - (EN50155: Heat, Cold, Insulation, Vibration and Shock)
  - EN60068-2-1,EN60068-2-2, EN60068-2-30

- Compliance EMC: EN50121-3-2:2017
  - Railway Applications - Electromagnetic Compatibility - Part3-2
- Compliance Shock and Vibration : EN 61373:2017
  - Railway Applications
  - Rolling stock equipment – Shock and Vibration Tests. Category 1 ; Class B
- Compliance Fire Protection - EN45545-2:2015 Railway Applications -
  - Fire Protection on Railway Vehicles - Part 2
  - Requirements for Fire Behavior of Materials and Components.
- Compliance to Ingress Protection IEC 60529IP54
- Compliance to EN55032 Electromagnetic Compatibility of Multimedia Equipment
  - Support MVB, RS-422,RS-485
  - Support reception of GPS, GLONASS, QZSS, BeiDou
  - Support Software Defines LAN

3.19 Future LTE-R Test Setup for KAVACH

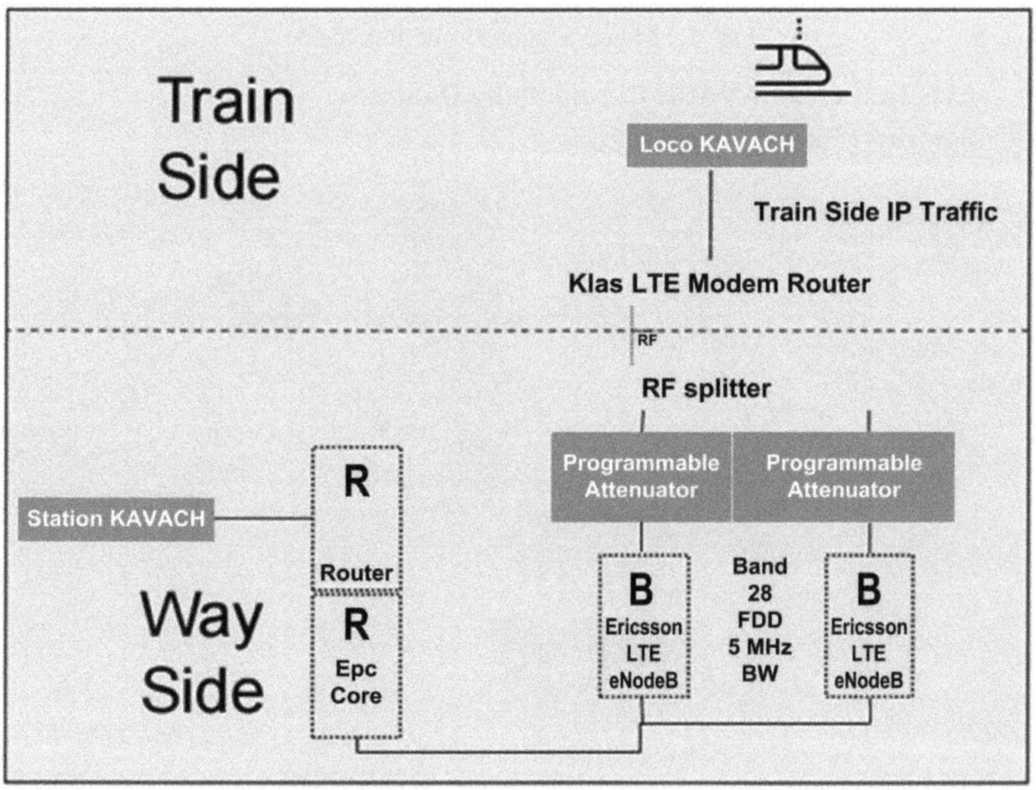

Fig. 3.16 Multiple Access Scheme

LTE Test Lab-KAVACH Connectivity Diagram

Scenario-1 Basic Communication b/w Loco KAVACH to Stationary KAVACH

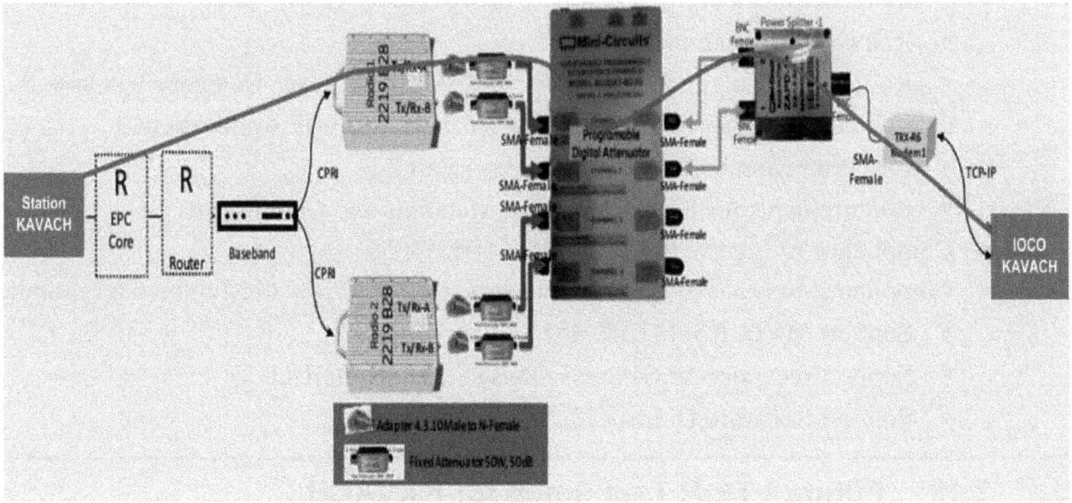

Fig. 3.17 Loco/Station Communications

LTE Test Lab-KAVACH Connectivity Diagram

Scenario-2 Radio/Sector Handover

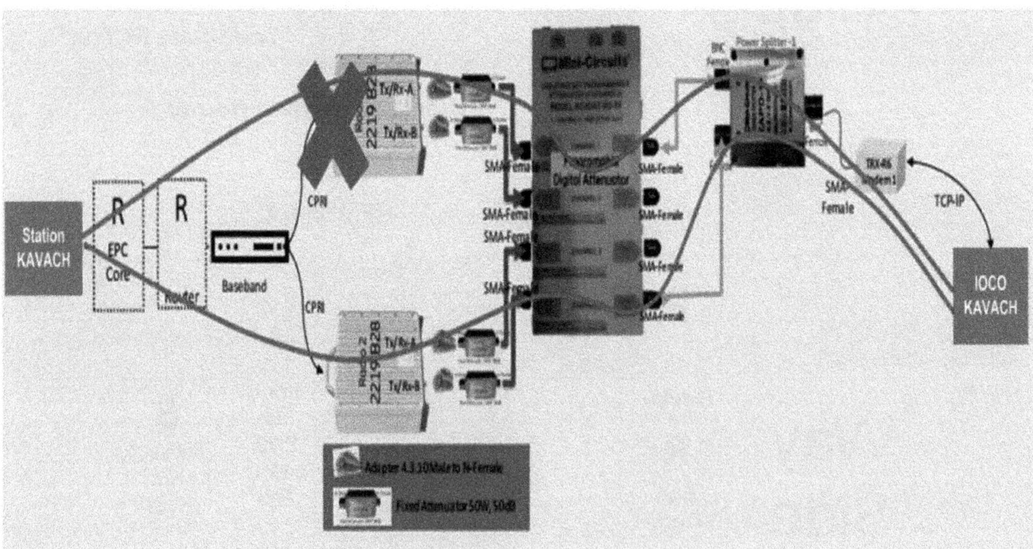

Fig.3.18 Station to Loco Communication

LTE Test Lab-KAVACH Connectivity Diagram

Scenario-3 Loco KAVACH communicating via both Modems
- Active/Active

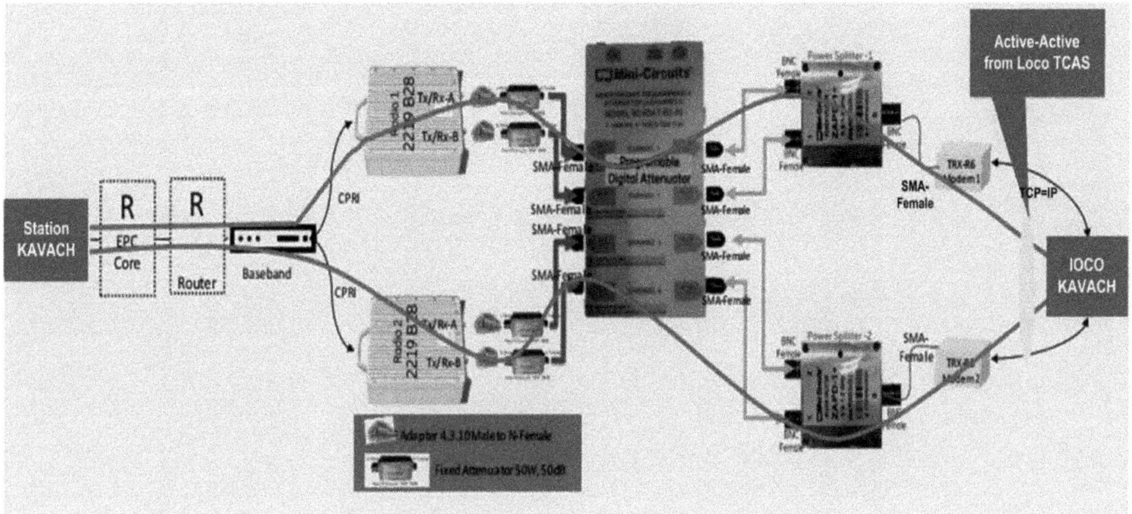

Fig. 3.19 Loco to Modems Communications

LTE Test Lab-KAVACH Connectivity Diagram

Scenario-4 Loco to Loco Communication

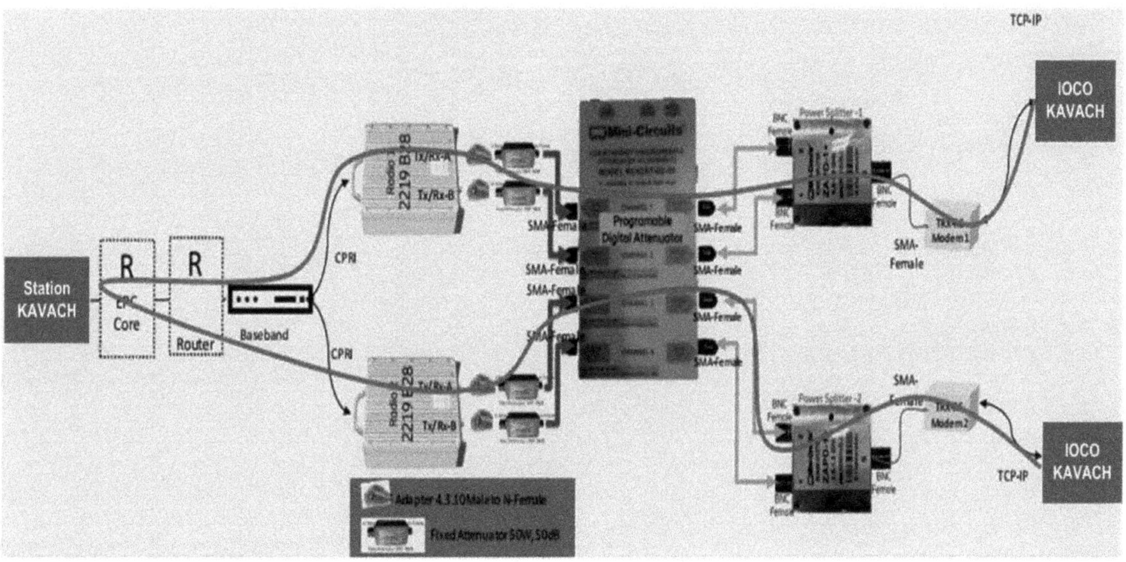

Fig. 3.20 Loco to Loco Communication

CHAPTER

4 Communication Techniques in KAVACH

4.0 Introduction

Communication plays a vital role in the working of KAVACH. The following three techniques shall be used in KAVACH for communication:

(a) Radio Communication

(b) GSM /GPRS Communication

(c) GPS/GNSS Communication

Radio communication is used for communication between Stationary KAVACH and Loco KAVACH as well as Loco KAVACH and Loco KAVACH. GSM/GPRS communication techniques are used for transfer of Authentication keys to Stationary KAVACH and Loco KAVACH through a Key Management System (KMS) Server for establishing radio communication between them. Fault messages from Stationary KAVACH and Loco KAVACH are also communicated to the Network Monitoring System (NMS) through respective GSM interfaces.GPS and GNSS are used for time synchronization purposes.

4.1 Radio Communication

The communication between Stationary KAVACH and Loco KAVACH as well as Loco KAVACH and Loco KAVACH shall be Full Duplex Radio Communication through Time Division Multiple Access (TDMA)/Frequency Division Multiple Access (FDMA) scheme under Radio communication protocol for Loco and Stationary KAVACH sub-systems.

4.2 Initiating Radio Communication

- As soon as a train enters a KAVACH territory by reading Signal/Signal Approach/ Normal/TIN Discrimination RFID tag, the Loco KAVACH unit shall commence transmission of the Radio message packet to Stationary KAVACH unit.

- Before establishing the direction, it shall send Absolute location and TIN as zero along with the last Signal/Signal Approach/Normal/TIN Discrimination RFID tag read, in this packet.
- After establishing the direction, Loco KAVACH unit shall commence transmission of absolute location and TIN as per the direction on the basis of data read from RFID Tag.
- The Stationary KAVACH unit shall transmit the Radio message packet to Loco KAVACH unit, if it receives the radio packet from Loco KAVACH unit with valid absolute location and direction.
- Whenever Loco KAVACH unit establishes communication with a new Stationary unit for the first time, it shall do so by Random Access Method within the time slots reserved for this purpose on frequency f_0.

4.3 Establishing Radio Communication

- On recognition of radio message from a fresh loco pertaining to the territory of Stationary KAVACH unit, the Stationary KAVACH unit shall allocate a Timeslot and Frequency Channel pair (e.g. f_1& f_2) for Communication with that particular Loco unit.
- Stationary and Loco KAVACH timeslots and frequency channel pairs shall be approved by User Railway. The frequency pairs allotted for two adjacent stations shall be different.
- Once, Loco KAVACH unit receives regular packet from Stationary KAVACH unit, Loco KAVACH unit shall continue further transmission in the frequency channel and time slot as specified in the Access Authority packet sent by Stationary KAVACH unit.
- For communication among different Loco KAVACH units in block section, in station area and in emergency situations (SoS, head-on, rear-end collisions), Loco KAVACH shall use transmit frequency (f_0) in its designated timeslot.

4.4 KAVACH Multiple Access Scheme & Radio Communication Protocol

- Full Duplex UHF Radio Communication is used for transfer of information between Stationary and Locomotive units. A specific frequency pair is allotted to a station for communication between Stationary KAVACH and Loco in station area. For example, if a Stationary Unit transmits on a frequency channel f1 and receive on f2, then Locomotive Units in its jurisdiction shall transmit on frequency channel f2 and receive on f1. The frequency pairs allotted for two adjacent stations shall be different. Stationary KAVACH System shall transmit the Radio packet in its designated time slot. Loco KAVACH System shall transmit the radio packet in its designated time slot (different from Stationary KAVACH TDMA slot) and designated frequency channel received from stationary KAVACH Unit.

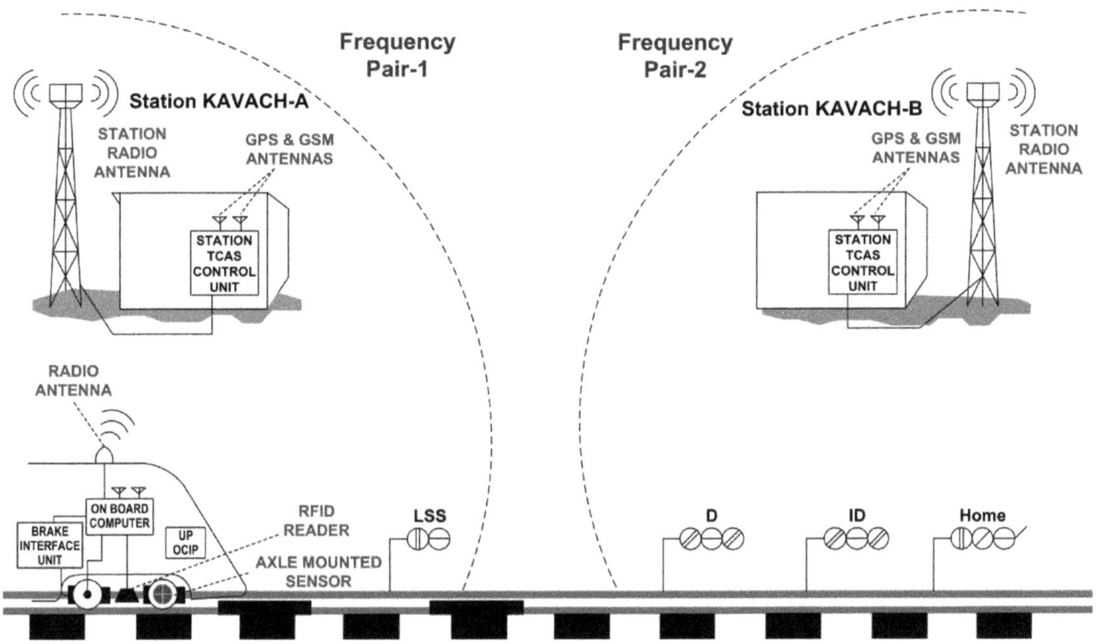

Fig. 4.1 Radio communication arrangement in Absolute Block Section

Each Multiple Access TDMA/FDMA frame cycle shall be of 2000 milli seconds. The frequency range of communication is 406 MHz to 470 MHz. The optimized frame cycle structure of the TDMA/FDMA scheme for Stationary KAVACH is shown in Figure 2. Suppose there are n stations ranging from1, 2, 3, 4……n. which comes under KAVACH territory.

- Figure shows pair of frequencies f_{S1}-f_{M1}, f_{S2}-f_{M2}.......... f_{Sn}- f_{Mn} allotted for stationery KAVACH units of adjacent stations where $f_{S1}$is the frequency on which Station1 KAVACH transmits and f_{M1} is the frequency on which it receives and soon.
- The frame cycle is divided into basic 78 time slot position markers (position nos. 1 to 78) each of width 352 bits (18.33m-sec).
- Out of 78 time slots, 50 Time slots shall be used for communication between Stationary KAVACH and Loco KAVACH. These timeslots have been marked as M-1 to M-50.
- These 50 time slots are distributed among adjacent stations sequentially till the last time slot, thereafter the sequence is repeated. Each station is allotted a block consisting of a number of timeslots in 2000 milli seconds frame. A particular Station KAVACH communicates with all the Loco KAVACH units in its jurisdiction in these allocated time slots in a specific frequency pair (say f_{S1}-$f_{M1}$for Station 1). The number of time slots allotted to a station depends upon the size of the station. Smaller stations like 3 or 4 line stations shall require less number of time slots, while stations with bigger yards may require more number of timeslots.

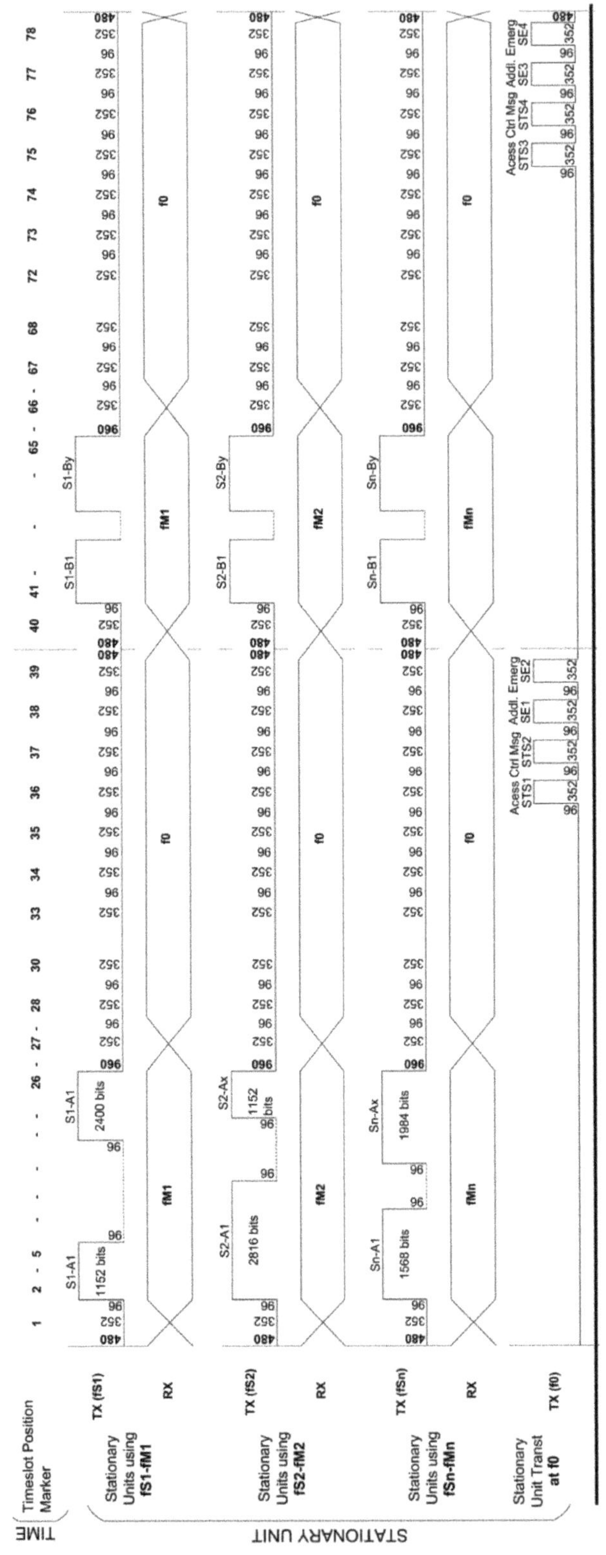

Fig. 4.2 Multiple Access TDMA/FDMA scheme for Stationary KAVACH

- The remaining Time slots other than the 50 mentioned as above shall be used by Stationary KAVACH and Loco KAVACH for communication in frequency f0 for various other purposes.
- Some specific Time slots in f0 shall be used by the Stationary KAVACH for broadcasting additional emergency (SoS) messages.
- Some other Time slots in f0 shall be used by the Stationary KAVACH for broadcasting Access Authority messages.
- In figure 3 frequencies f_{S1}-f_{M1}, f_{S2}-f_{M2}........... f_{Sn}-f_{Mn} allotted at adjacent stations for respective Loco KAVACH units where f_{M1} is the frequency on which LocoKAVACH unit in Station1 area transmits and f_{S1} is the frequency on which it receives and so on.
- On receipt of radio message from a fresh loco falling in the jurisdiction of Stationary KAVACH unit, the Stationary KAVACH unit shall decide the Timeslot in which that particular Loco unit shall communicate.
- Stationary KAVACH and Loco KAVACH shall use different timeslot(s) for communication. It means that Stationary KAVACH and Loco KAVACH shall use their respective timeslot(s) in the Multiple Access within their channel for the transmission.
- Although the frame cycles for Stationary KAVACH and Loco KAVACH are different at a particular station, Stationary KAVACH and Loco KAVACH time slots shall not overlap.
- The Loco time slots for the same station shall not be adjacent to each other (minimum one time slot gap shall be kept as per TAN5001).
- In block section, Loco KAVACH System shall switch its Tx frequency to f_0 and shall transmit the radio packet in some specific time slots for communicating with Stationary KAVACH.
- In emergency situation, Locomotive shall additionally transmit on f_0 in a particular timeslot and tune back to receive on f_0. Transmission of SoS messages by Locomotive on f_0 in case of emergency situation shall enable other trains in vicinity to get the immediate message without routing through Stationary Unit.
- Loco KAVACH unit while transmitting on frequency f_0 shall not interfere with the portion of the frame cycle nominated for transmission of Stationary KAVACH units.

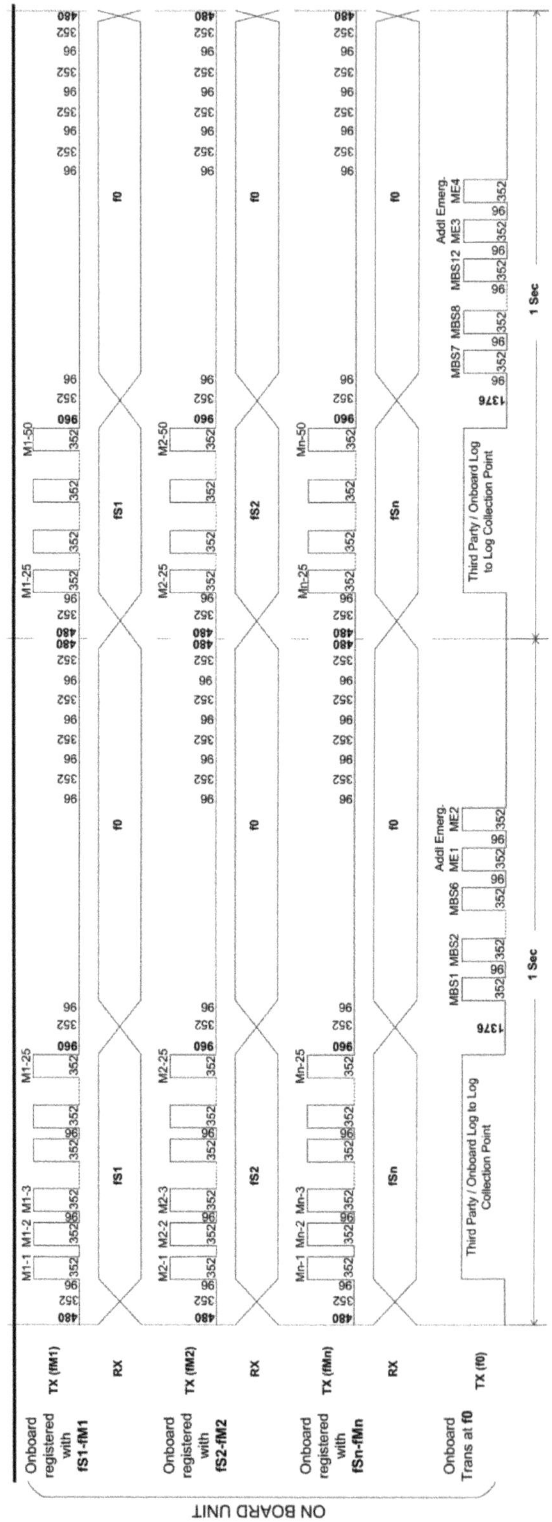

Fig. 4.3 Multiple Access TDMA/FDMA scheme for Loco KAVACH

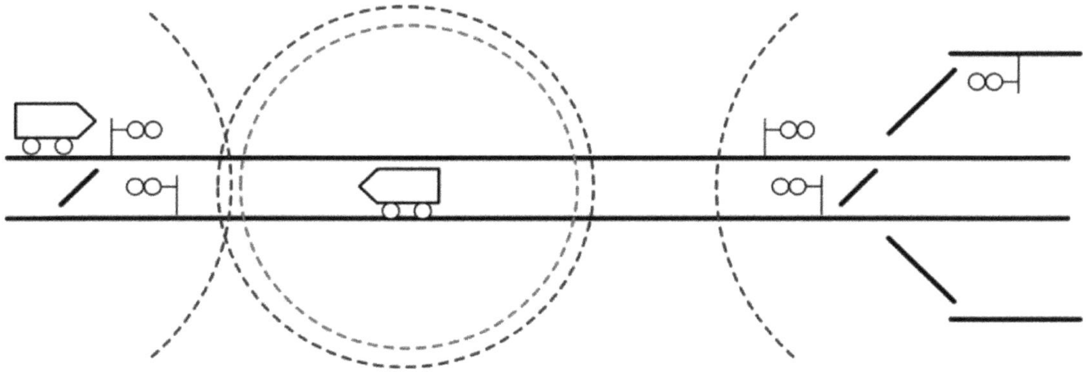

Fig. 4.4 Communication by Loco KAVACH in Absolute Block Section

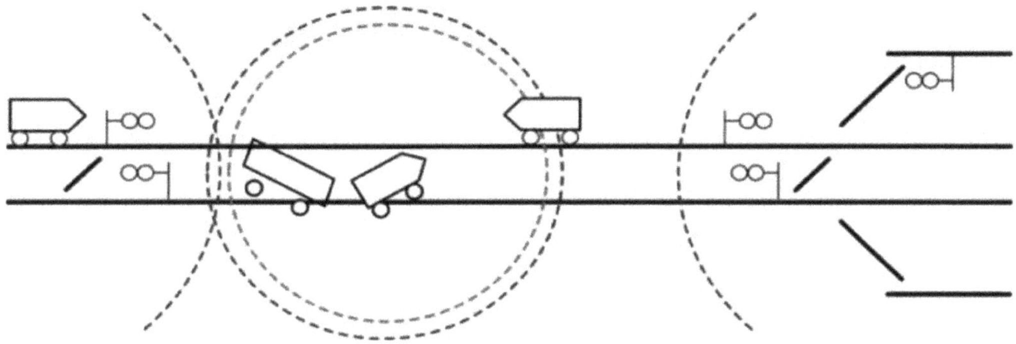

Fig. 4.5 Communication by Loco KAVACH in emergency situation in Absolute Block Section

4.5 Communication in Overlap Zone

The **communication mandatory zone** for a stationary KAVACH unit shall include at least two RFID tags prior to a distance of 1 km from first approaching signal of the respective stationary KAVACH unit in Absolute Block Section.

In case there is no gap between Communication Mandatory Zones of two adjacent Station Interlocking, there shall have to be overlapping zone in which Loco KAVACH Unit would receive messages from both the Stationary KAVACH Units. On receiving **Access Authority** from Stationary KAVACH Unit on approach, Loco KAVACH Unit shall relinquish communication with preceding Stationary KAVACH Unit and start communication with approaching Stationary KAVACH Unit.

Unlike absolute block section, Radio communication is mandatory for the entire automatic Block section. RF overlap in auto section is required to be achieved by using two frequency pairs in adjacent stations.

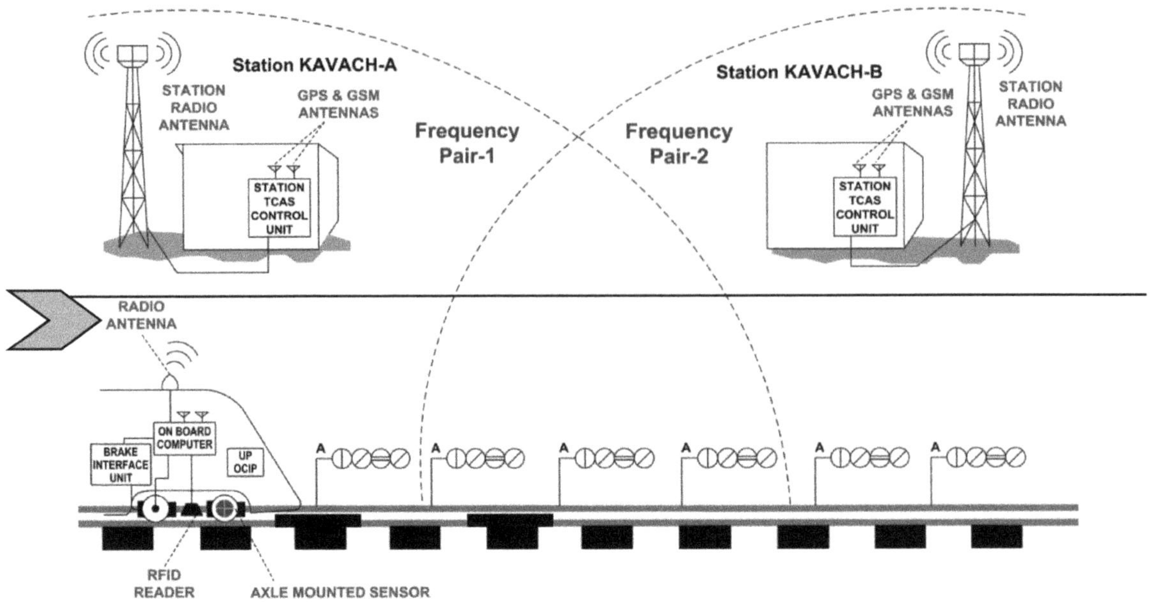

Fig. 4.6 Radio Communication arrangement in overlap zone in Automatic Block Section

4.6 Terminating a Radio Communication

Stationary KAVACH shall stop communication with Loco KAVACH unit, when one of the following conditions occurs:

- When the Loco moves beyond the Last Stop Signal of the Stationary KAVACH Unit as per Boundary configured (typically 1.5 km from Last Stop Signal of Stationary KAVACH).
- If the direction is invalid in Standby or Staff Responsible mode.
- No radio packet is received from Loco KAVACH for more than 60 cycles.
- In the event of failure of both – GPS/GNSS and RTC,

If no communication is received from a registered loco KAVACH unit in communication mandatory zone for continuous 2 minutes, the same shall be deregistered by the stationary KAVACH unit.

4.7 GSM & GPRS Communication

The GSM/GPRS communication is used for following purposes in KAVACH:

Communicating Fault messages to NMS

The exceptional fault/critical messages logged by Stationary KAVACH and Loco KAVACH unit are transmitted to NMS through respective GSM interfaces provided in hot standby mode. Stationary KAVACH shall log fault messages and transmit to Network Monitoring System in following cases:

- Radio Communication failure longer than applicable time-out (30 seconds).
- Failure of only one radio out of two radios in hot standby in Loco KAVACH.
- CPU time and GPS time out of synchronization.
- Failure of both – GPS/GNSS and Real Time Clock (RTC).
- Active Cab/Desk DMI communication failure in Loco KAVACH.

Fig. 4.7

Loco KAVACH shall log fault messages and transmit to Network Monitoting System in following cases:

- RFID reader failure
- In Post Trip (PT) Mode or Isolation (IS) Mode.
- On receiving Missing RFID Tag information.
- Radio Communication failure longer than applicable time-out (30 seconds).
- Failure of only one radio out of two radios in hot standby.
- Failure of both GPS/GNSS and Real Time Clock (RTC).
- Active Cab/Desk DMI communication failure.

Transmission of Authentication keys for radio communication between Stationary KAVACH and Loco KAVACH

When the Stationary KAVACH unit communicating the safety related data with Loco KAVACH, it shall verify that communication is established with an authorized Loco KAVACH unit and vice versa. Consequently the authenticity and integrity of any information exchanged between Loco KAVACH and Stationary KAVACH unit shall also be verified. In order to establish secured communication between Stationary KAVACH and Loco KAVACH units, Authentication keys are transmitted to all Stationary KAVACH and Loco KAVACH units through a Key Management System (KMS) using GSM communication. Stationary or Loco KAVACH units shall send the request to KMS for Authentication key by using GPRS. Authentication key will have certain valid time period for its usage. The communication between Stationary KAVACH and Loco KAVACHunits is encrypted through Authentication key and thus the authenticity and integrity of transmitted information is ensured.

Note: All messages in time slots reserved for emergency messages and Loco-to-Loco direct communication are not subjected to aforesaid procedure of Identification and Authentication dialogue and are non-SIL-4 due to nature of the information conveyed by such messages.

4.8 Key Management System (KMS) Server

Key Management Server

Key Management Server (KMS) is designed to share secret Authentication keys to Stationary and Loco KAVACH units. Message origin authentication/message integrity is a safety procedure ensuring the integrity and authenticity of messages during Radio communication. These Authentication keys are used to protect the messages against modification and to ensure that no one can masquerade as the originator of the message. They are shared only after authenticating the Loco KAVACH through OTP. KMS is deployed on a secured server on the Internet. All the Loco KAVACH units request for the Authentication keys periodically and use the same for Radio Communication.GSM modules residing in the Event Logger module are used to connect Loco KAVACH to GPRS (Internet).

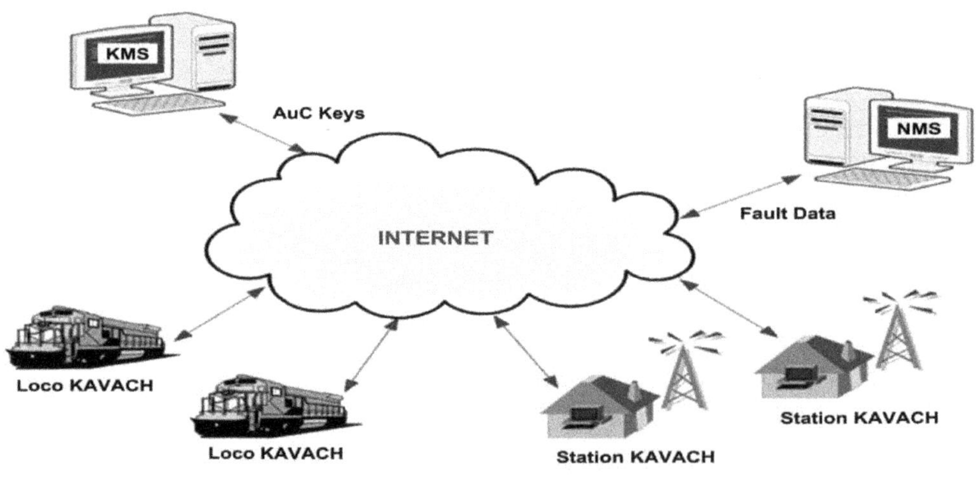

Fig. 4.8

- KMS is responsible for distribution of authentication keys to KAVACH systems for radio security.
- Communication between Stationary and Loco KAVACH Units is based on AES-128 encryption.

- GSM communication is used to transfer Keys to Loco and Stationary KAVACH.
- A central KMS server is hosted at Head Quarters to communicate authentication keys to Loco KAVACH and Stationary KAVACH.
- Single KMS is sufficient for all Railways. All the stationary KAVACH and Loco KAVACH IDs and SIM card numbers are to be provided to Railtel, to update the same in the KMS.
- Communication between KMS and KAVACH is on GPRS and currently being upgraded to LTE.

4.9 Key Definition & Type of Keys

Key is defined as a sequence of 128 bits. The following table summarizes the key types and usage of each type of key.

4.9.1 Authentication Key (K_A)

Authentication key is used for session key derivation in order to establish a safe connection between Stationary & Loco KAVACH units. This key will be communicated by Key Management System (KMS) to all Loco and Stationary KAVACH units and it will have valid time period for its usage. It can be changed from time to time. Key Management System would be centralized for Indian Railways. KMS shall use OTP (One Time Password) over SMS technique, through GSM, to authenticate the KAVACH systems before sharing the Authentication Keys. KAVACH units shall store the Authentication Keys.

4.9.2 Session Key (K_S)

- Session key is used for protection of data transfer between Stationary & Loco KAVACH units. This key will be computed by Loco KAVACH and Stationary KAVACH units at the time of establishment of communication session between Stationary KAVACH unit and Loco KAVACH Unit.
- On entering into "Communication mandatory Zone" in vicinity of Stationary KAVACH territory, Loco KAVACH unit sends the Access Request packet to Stationary KAVACH system in "f0" frequency.
- Stationary KAVACH unit transmits the Access Authority message with Message Authentication Code (MAC).
- Loco KAVACH unit receives the Access Authority message packet with MAC and compute the session key KS.
- The MAC will be computed for the Access Authority message packet and verified with received MAC. If the computed MAC matches with the received MAC, Loco starts communicating the regular packet.

4.10 Remote Interface Unit (RIU)

- Remote Interface Unit (RIU) shall be used where remote signalling functions are required to be fetched to a nearby Stationary KAVACH unit.
- RIU shall have provision for interfacing with signalling inputs in fail-safe manner.
- RIU shall utilise line modems for communicating with stationary KAVACH unit.
- A single RIU shall be capable of handling at least 32 field inputs
- OFC or Quad cable, ring network shall be formed to increase availability of the network.
- A single RIU shall be capable of communicating with two adjacent RIU units so that the operations are not affected in case of communication link failure on one side only (2oo2 architecture).

4.11 Radio Communication Arrangement

Station KAVACH and Loco KAVACH communicates with each other through Radio modems.

4.12 Radio Modems

- The Radio with a transmission power of 10W is capable of communicating with the other KAVACH within the radius of 5 KM at- least.
- Antenna for station/IBS/mid-section interlocked Gate unit will be Combination of vertically polarized Omni-directional antenna.

The tower for Radio communication will be of 40 meter height

- Stationary KAVACH and Loco KAVACH units communicates with each other for every 2 seconds.
- Pair 1 : Station Tx - 456.8 MHz, Rx - 416.8 MHz
- Pair 2 : Station Tx - 466.8 MHz, Rx - 426.8 MHz
- Common Frequency, f_0 - 441.8 MHz

4.13 GPS/GNSS Communication

- Time synchronization is very important for seamless data communication among various KAVACH units. Absence of time synchronization often results in error/faults. Multiple Access TDMA scheme is used in KAVACH which requires unified time synchronization. We do not have a Universal Time clock in India, hence we use GPS/GNSS Clock to correct the Real Time Clock (RTC) of any system.
- Station/LC/IBS Vital Computer shall have Real Time Clock (RTC) synchronization facility with GPS/GNSS clock to synchronize with other KAVACH systems in hot standby manner.

- GPS antenna is installed at the stationary KAVACH unit for time synchronization. Two separate antennas for GPS shall be installed using diverse path for high availability. Diverse make of GPS are preferable to avoid common cause failures.
- The Station/LC/IBS Vital Computer RTC should be synchronized with the RTC of all Loco KAVACH Vital Computers and other KAVACH systems. The RTC of all the KAVACH units are set with time reference to GPS/GNSS.
- The factors affecting time synchronization are temperature, inferior quality clocks, resource constraints, high density and node failures. When RTC of any Station/LC/IBS KAVACH or Loco KAVACH is not matching with the RTC of other KAVACH units, it synchronizes with GPS/GNSS clock.
- When CPU time of Station/LC/IBS KAVACH or Loco KAVACH and GPS time are not synchronized, the time reference shall change to other GPS.
- If both GPS are failed, then the system shall work on CPU time for 30 minutes until the situation is stabilized. If there is no stability after GPS time-out, the Loco shall transit out of Full Supervision Mode.

CHAPTER

5 Network Monitoring System (NMS)

5.0 Introduction

Network Monitoring System (NMS) shall be is provided over OFC Network (E1 interface) for centralized monitoring of KAVACH equipped Trains and Stations within the network. It is used for troubleshooting of error events, off line simulation, real time monitoring of loco KAVACH etc. Stationary KAVACH and Loco KAVACH transmit exceptional fault/critical messages to NMS. A Central Server is present in the Division Control Room. All relay information and radio packets exchanged between Station and Loco are logged in Central Server and accessed through NMS.

5.1 NMS Hardware Requirements

The hardware requirements of NMS are as follows:

- 2 Nos. Servers for redundancy
- (NMS data shall be logged onto each server) and Shared Firewall on both servers
- Keyboard and Mouse

55" 4K/UHD Professional TV. (If one screen is insufficient for display, separate screen shall be provided with seamless integration to enable complete view of KAVACH territory).

5.2 Stationary KAVACH to NMS & Stationary KAVACH to Stationary KAVACH Communication on E1 Interface

- KAVACH NMS Network is built on E1 interface which is provided by the Indian Railways. Centralized monitoring of a group of stations is achieved by collecting signal aspects, track occupancy, loco absolute position etc., from each of the Stationary KAVACH unit within the network. Network monitoring is limited to its own network. Stationary KAVACH units communicate with NMS unit using the predefined packets.

- Using E1 interface, each Stationary KAVACH unit is connected to adjacent stationary KAVACH unit/Network Monitoring System to form a network, as shown in Figure 6.1. Using Ethernet protocol over this network, Stationary KAVACH units will exchange Stationary-Stationary communication packets with adjacent Stationary KAVACH units and NMS.
- Number of Stationary KAVACH units in one E1 ring shall be limited to 5-7. The IB huts are connected to the nearest Station in T network.

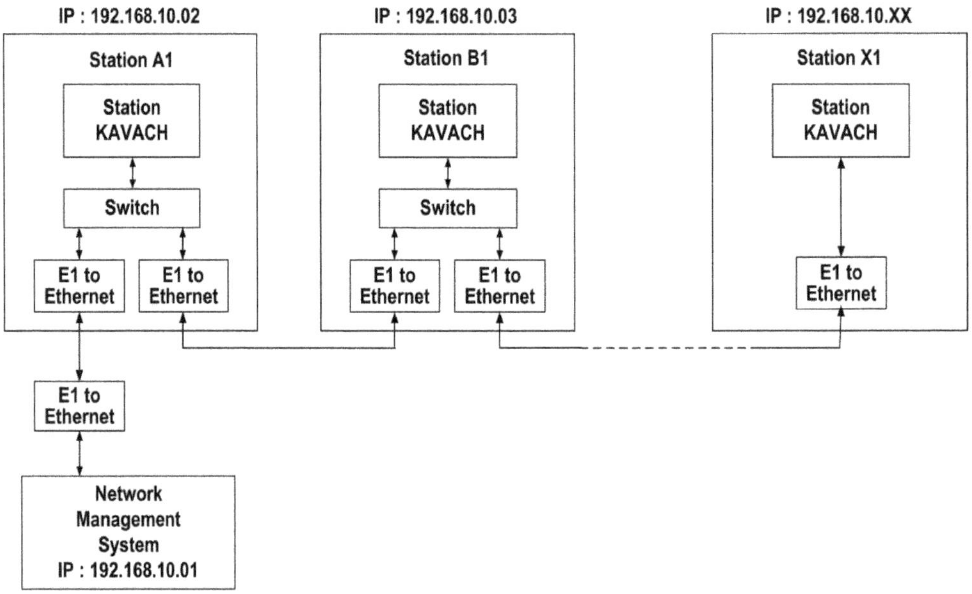

Fig. 5.1 Network Monitoring System (NMS) connectivity diagram

Stationary KAVACH shall log fault messages and transmit to Network Monitoring System through Ethernet or GSM interfaces in the following cases:

- In the event of a Radio Communication failure longer than applicable time-out.
- In the event of a failure of only one radio of Loco KAVACH, out of two radios in hot standby.
- If the incremental difference between CPU and GPS time is not matching and the difference between two GPS is greater than the frame interval.
- In the event of failure of both – GPS/GNSS and Real Time Clock.
- In the event of Active Cab/Desk DMI communication failure.

5.3 Loco KAVACH to NMS Communication on GSM Interface

The communication between Loco KAVACH and Network Monitoring System (NMS) shall be through two GSM interfaces provided in Loco KAVACH Vital

Computer. The exceptional fault/critical messages logged by Loco KAVACH unit are transmitted to NMS through two GSM interfaces provided in hot standby mode.

Loco KAVACH logs fault messages and transmits to Network Monitoring System through GSM interfaces in the following cases:

- In the event of a RFID reader failure (any one or both readers).
- To indicate the Loco KAVACH mode to NMS through Stationary KAVACH Unit on encountering any tag (excluding LC Gate Tag), when Loco KAVACH unit is in Post Trip (PT) Mode or Isolation (IS) Mode.
- In case of missing RFID Tag information received from Stationary KAVACH. This in turn send an SMS alert to the Maintenance Staff.
- Information to NMS if the turn out tag is over read by Loco KAVACH unit.
- In the event of a Radio Communication failure longer than applicable time-out (30 Seconds).
- In the event of a failure of only one radio out of two radios in hot standby.
- In the event of failure of both GPS/GNSS and Real Time Clock (RTC).
- In the event of Active Cab/Desk DMI communication failure.

5.4 Salient Feature of NMS

- Real time display of train movement on NMS monitors.
- Offline display of train movement on NMS monitors at Normal, 2x, 4x, 8x, 16x & 32x speed.
- Display of Data log in Spreadsheet (Excel Format).
- Display of "Permitted Speed+ Current Speed Vs Location" and "Permitted Speed+ Current Speed V/s Time" including information whether brake command is applied by KAVACH or not.
- Ability to watch the NMS at Distant Location through Internet (Password protected)
- Ability to extract offline data log through NMS
- Generation of Exception Reports - Loco KAVACH Unit-wise, Stationary KAVACH Unit-wise, RFID Tag Set wise.
- Prompt for missing one of the two RFID tags of same set.
- Prompt for missing both RFID tags of same set.
- Prompt for missing communication packets overall below a set level (say 20%) for moving train in Communication mandatory zone.
- SMS Alert for repeated same RFID tag missing events in Full Supervision Mode.

- SMS Alert for any brake application command by KAVACH Loco forcing train to bring to dead stop in Full Supervision Mode.
- SMS Alert for SPAD.
- SMS Alert for SPAD Prevention by KAVACH.
- Capturing of speed restrictions imposed by TSR Management System from Stationary KAVACH after every update.
- Capturing health status and event log from Loco KAVACH if LTE is available.

5.5 NMS OPERATION

5.5.1 NMS Login

Application deployed on a web server which can be accessed using any web browser which is in the network of NMS (192.168.10.0/24). The IP address for the NMS server is 192.168.10.1

Basic procedure for connecting the NMS server is given below :-

1. Open the web browser in any PC which is connected in the NMS network
2. Enter the address in the browser as 192.168.10.1/nms for accessing the login page as shown in figure 5.2.
3. Enter the credential Username - iriset , Password - XXXX and click on the login button

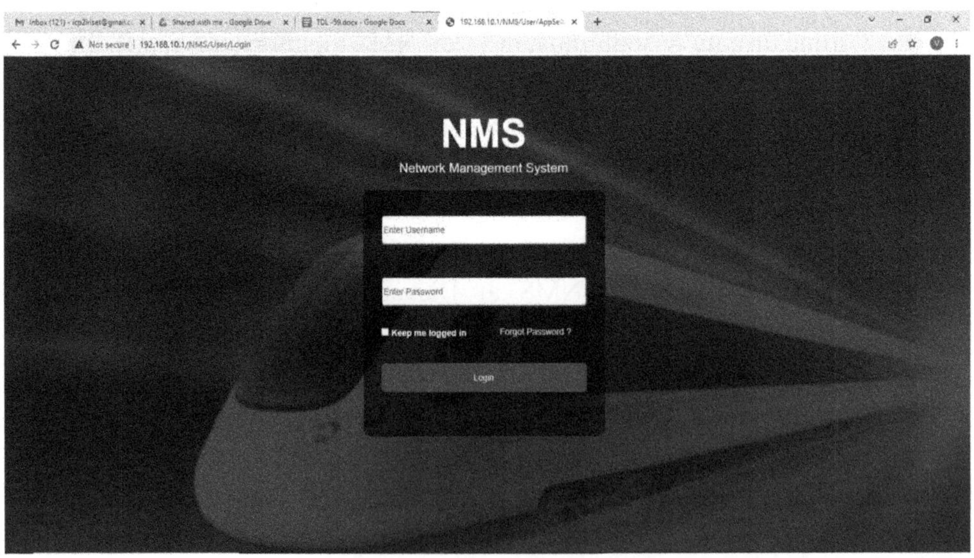

Fig. 5.2

4. After the login button is pressed an app selection page will open with all four option which is available on NMS server as shown in figure 5.3.

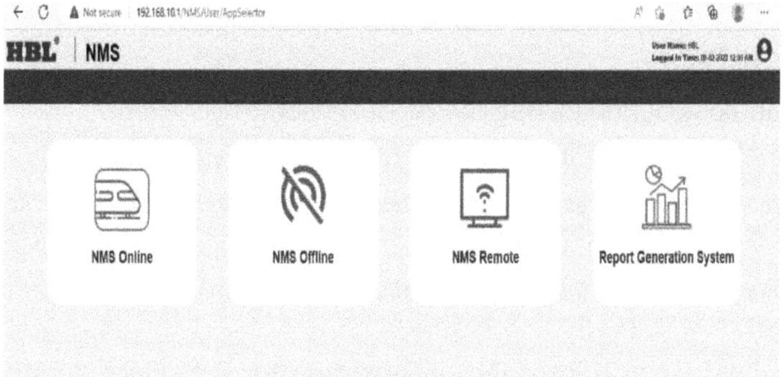

Fig 5.3

Now as per requirement above particular object has to be selected.

5. NMS online is selected then the page with selection of yard will come where depending upon the user choice any or all the stations can be selected. After selection open button is clicked.
6. After selection of yard on station selector page a complete layout as per selection will be visible with station name in white on the screen
7. Now ethernet connectivity is to be established by using Menu→ethernet config.→connect.After this the whole layout will be connected with NMS as per the network condition . If in link then all station name will be green
8. Now whatever the movement of loco in the connected section will be visible on the NMS screen with loco icon & loco with loco Id. Loco will display 6 values like Loco ID, speed,Absolute location, RFID, Length of train & mode of loco as shown in the Fig. 5.4.

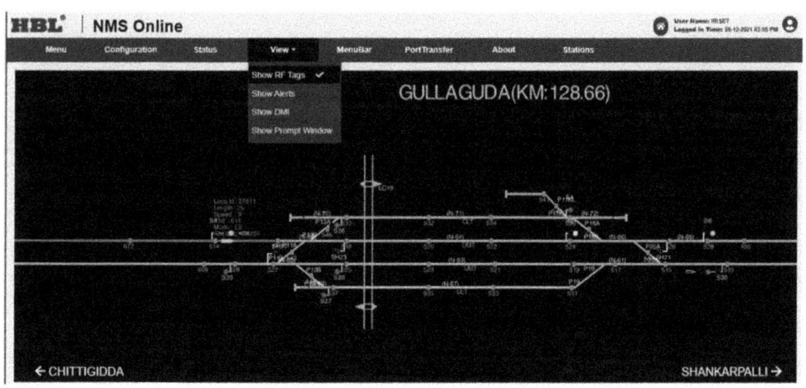

Fig. 5.4

9. Other details related to online simulation can be accessed using different option available in on home screen (dashboard) like status(Relay & loco) & can view RF tags, Alerts, DMI & prompts.

5.5.2 Reports

Reports can be generated using the report generating system.

Specific steps required for generating reports are given below:-

1. Report generation system is selected from app selection screen as shown in Fig. 5.5
2. To generate a report for a particular time, From & To time is and apply button is clicked
3. If the database will have the information for the selected date a prompt message will confirm for the same

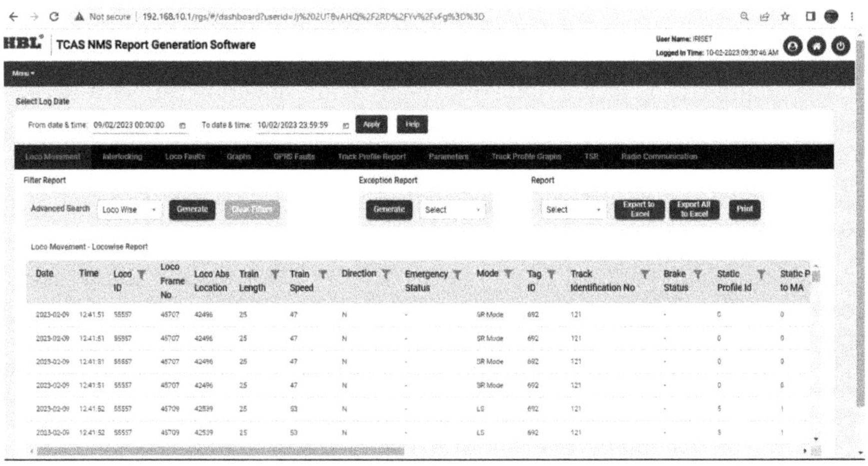

Fig. 5.5

4. Now report for various events like, Loco movement, Interlocking, Loco faults, Graphs, GPRS faults, Track Profile report, Parameters & Radio Communication can be obtained /populated using the appropriate selections as shown in Fig 5.4. One has to generate the report every time on selection of some parameters afterward a report will be generated.
5. Report generated can be exported also on suitable format.

5.5.3 Playback

Playback NMS will have a provision for offline NMS, one of the feature of NMS.

Application will display list of stations with station name and code from the Database file (Fig 5.6).

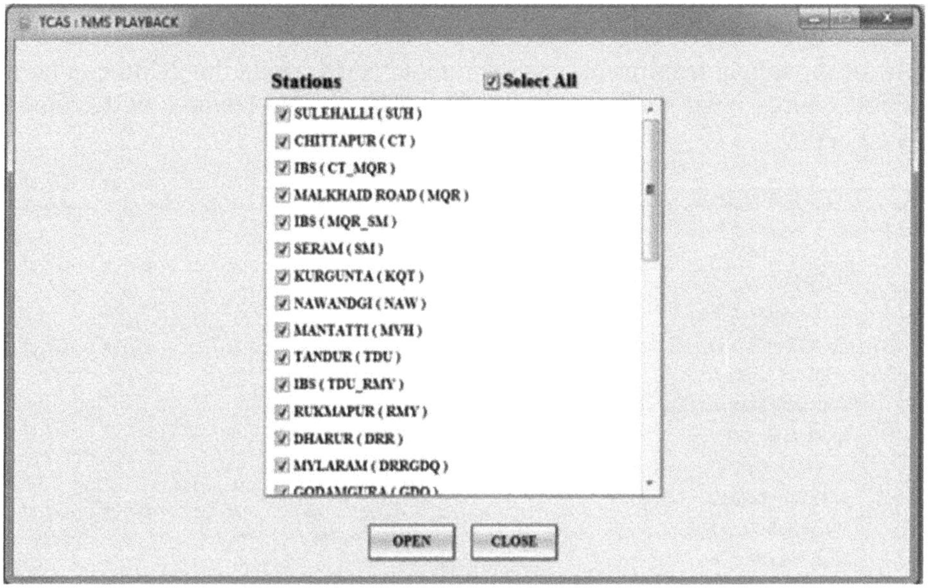

Fig. 5.6 Playback NMS home screen

- Either any or all the stations of the network can be selected by using GUI based applications.
- After selection of a particular yard or whole playback option is provided
- Based upon the selection of the database file (bin file) from the log files which have been saved in the database, the replay NMS tool plays the NMS for the given selection
- Even the specific station or specific loco movement can be replayed using the selection, otherwise it will play the whole network events for the selected date
- Replay NMS has provision for player control having pause & stop the content along with time slider
- Application will display different speed options for the playback which is selectable from 1x, 2x, 4x, 8x, 16x and 32x.
- Application also has the provision for tracking a particular loco using selection
- Application has provision for exporting the content which is played on replay NMS.
- Application will display playing time on screen.

5.6 Remote NMS

One of the salient features of NMS is remote NMS, using this NMS can be watched at remote places other than server location. The salient features of Remote NMS are listed below:

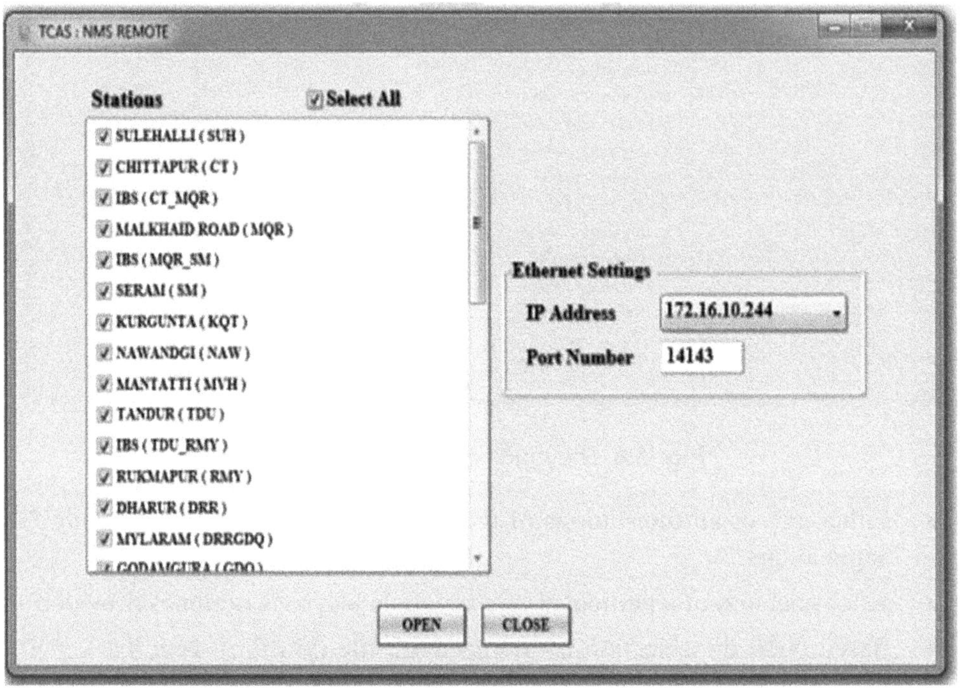

Fig. 5.7 Remote NMS screen

- Application will display a list of stations with station name & code which can be selected as in case of NMS like either or all at a time.
- Application will display a list of IPs available in the PC from which user can select one IP at a time for connecting NMS (Fig. 5.7).
- Application will display a list of stations with station name & code which can be selected as in case of NMS like either or all at a time.
- Application will display a list of IPs available in the PC from which user can select one IP at a time for connecting NMS (Fig. 5.7).
- User has to enter an available Port number for NMS Communication.
- After entering the IP & port details using the OPEN button Remote NMS main screen will be opened after validating the input provided by the user.

- Remote NMS main screen will have a login option .Afterclicking on login option a pop window will open, upon entering the credentials provided for remote NMS connect button is used to connect to Remote NMS
- After successful validation of credentials Remote NMS connected screen will appear which will work like normal NMS with login remote user name on GUI, else it will show appropriate error messages on the screen.
- Remote NMS allows a minimum 5 nos. of remote users simultaneously.

CHAPTER

6 Remote Interface Unit (RIU) used in KAVACH

6.0 Introduction

Refers to requirement of Remote interface unit which can be installed in auto, IB or gate signal etc. Requirements for adapting KAVACH to Auto Signaling Territory. In an Auto Section, signals are set up and cleared autonomously by track occupancy detection systems.

6.1 Remote Interface Unit (RIU)

The following criteria shall be adopted for provision of **RIU**:

1. In case where concerned repeaters for signals of End Cabins/Distributed Interlocking/LC Gate/IB are not available at station, but radio signal strength of station KAVACH tower is adequate at said End Cabins/Distributed Interlocking/LC Gate/IB location, separate KAVACH RIU unit shall be installed at such location to bring the aspects over OFC to the nearby station whose radio tower is to be used. OFC Cable from Station KAVACH to RIU (at LC/IB/cabin) in diverse path has to be laid to increase the availability of the network.
2. In case the signal strength is not adequate at locations such as mid-section LC Gate/IB, separate Station KAVACH unit and tower to be installed at the said LC/IB location
3. The locations of such installations shall be finalized based on the findings of RSSI signal strength survey,
4. As a thumb rule, RIU may be placed at all LC locations which are less than 5 km from station. This is because radio signal strength may be adequate at such locations. For LC gates, which are more than 5 kms from station and radio signal strength is weak, separate KAVACH unit and Tower may be installed.

5. In double distant territory, where distant or inner distant signal aspect is not available at station, the same shall be extended to approaching station by laying of cable. Cable for this purpose will be supplied by the Railways. Alternately, the aspects of these permissive signals can be detected. The Railway will decide which option to exercise.
6. Suitable RIUs shall be provided in Automatic block section where Automatic Signalling is implemented by providing Signal Interlocking circuits in location boxes or Auto huts provided near signal posts to capture signal aspects. These RIUs shall be connected to a centralized Stationary KAVACH unit to ensure functioning of KAVACH in Automatic Signaling territory. Suitable power supply arrangement for RIUs is to be done by the concerned agency for this purpose.

6.2 Illustrative Diagrams

Automatic change of sequence of aspects behind the train in three-aspect is illustrated in the following Figure 6.1.

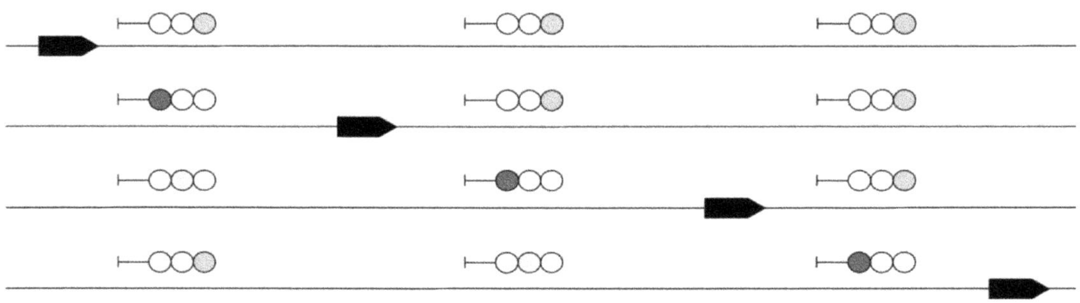

Fig. 6.1 Automatic change of sequence of aspects behind the train in Four Aspect Signalling Territory

- For adapting KAVACH to an Auto section, it is necessary to communicate the signal aspects or/and Track occupancy status to Stationary KAVACH, which then determines movement authority and communicates the same to the Loco KAVACH on radio.
- If the signal is at OFF and linked, loco shall transit to FS mode after passing the signal foot tag.
- If the signal is at OFF and not linked, loco shall transit to LS mode if SSP is available else transit to SR mode after passing the signal foot tag.
- If another KAVACH equipped loco is ahead of the loco, the signal aspect & MA shall not be linked to the rear loco in OS mode.

- In the case of single line working, KAVACH shall extend Movement Authority after ensuring the establishment of direction of traffic and all stop signals (if available) against the established direction shall be at ON.
- The modified automatic signal working as per General Rules shall also be applicable for KAVACH in both single line as well as double line working. IB working with ignoring signal foot tags at other signals to be ensured.

6.3 Interface Requirements

- Remote Interface unit shall have provision to interface only on redundant OFC Dark fiber for connectivity with Station/LC/IBS Vital Computer.
- Stationary KAVACH shall be able to connect minimum 06 Remote Interface Units in one direction (up & down towards one side of a station is considered as one direction). Stationary KAVACH shall be able to handle minimum 06 directions. Demonstration for various combinations of routes (Single/Double/ Triple/Quadruple) shall be carried through functional testing by defining appropriate test cases.
- Interface Unit. This failure shall be intimated to NMS through Stationary KAVACH. Remote Interface units shall be connected in ring fashion and they shall be able to communicate to stationary KAVACH in case of failure of Main Path. Stationary KAVACH shall be intimated about the failure of Main/Stand-by Path failure by the Remote
- Communication between RIU and Stationary KAVACH shall comply with EN 50159 (closed) Standards or equivalent standards\
- The cyclic redundancy check for the communication protocol shall be com- plying with the requirements of EN 50159 (Closed).
- Failure of primary or secondary communication links shall be logged in KAVACH NMS and also communicated through SMS.
- All the Hardware and Software employed in RIU shall be certified for EN 50126, EN 50128, EN 50129 and EN 50159 for SIL-4. Power Supply, Surge Protection and Ground

6.4 In Railway Electrification (RE) Area

Each RIU shall be fitted with a battery backed power supply with minimum 2 hour (Two Hour) with two chargers &, designed to receive an input voltage of 230V AC. The power demand of RIU shall be as low as possible. The responsibility of power supply compatibility is with OEM. The RIU shall operate without any problem, even if the Neutral to Earth voltage is as high as possible. Since, RIUs are connected through a power cable, where N-Earth and P-Earth and P-N insulation is prone to be low and hence, firms shall specify the recommended, minimum, and maximum permitted values of these with documental evidence. These parameters and min N-Earth voltage, min, max and recommended mains voltage shall be included in the pre-commissioning checklist

6.5 In Non Railway Electrification (Non-RE) RE Area

- The power supply system in Non-RE area should be provided with proper backup. The battery backup shall comply with EN 50121-4 or equivalent standards and RDSO/SPN/144 standards. The battery shall with stand for temperatures up to 60°C.
- The battery charger shall have provision to generate alarms for AC Input fail, Charger fail and low battery. These alarms shall be integrated with RIU. It shall be possible to generate SMS of these alarms from NMS.
- The surge protection arrangements for all the incoming and outgoing cables shall be provided as per RDSO latest applicable guidelines/specifications.
- The Grounding arrangements are required to be provided as per RDSO latest specifications and as per the equipment requirements.

6.6 Other Important Features

- The RIU shall have provision to Interface stationary KAVACH in two different paths for connecting to the Stationary KAVACH in the redundant manner
- The connection between RIU and the adjacent stationary KAVACH shall be through a dedicated dark fiber in redundant manner.
- The failure of the communication link with one RIU shall not result in failure of subsequent Remote Interface Units.
- Module wise health monitoring shall be made available at NMS through Stationary KAVACH.

6.7 Handing Over to Next Station Seamlessly

- The communication mandatory overlap zone where Loco is handed over and taken over shall consider minimum two Auto signals to be wired to both stationary KAVACH units. In the overlap section, both the stations will communicate with Loco KAVACH. Loco KAVACH should set up the communication based on the approaching Station ID. Once communication is set up, Loco KAVACH should ignore the radio packets coming from previous station.
- Signal whose aspects are to be shared by the handing over station and taking over station shall be wired to two Remote Interface Units. Two such signals per line

shall be wired to each stationary KAVACH through RIU inside the over-lap zone. Each Remote Interface Unit shall be wired to handing over station and taking over station respectively.

- In Loco Moving Direction, handing over station shall wire the taking over station's first auto signal as overlap and communicate it is signal aspect before taking over stationary KAVACH establishes communication with Loco.
- The Radio Signal Strength of both the handing over station and taking over station at the overlap zone shall be above -85dBm. If this signal strength is not achievable, then additional Stationary KAVACH shall be planned in the block section.
- The time slot distribution charts shall be prepared for the maximum design limit and functional testing for the same shall be carried out.

6.8 Handling of MA, SSP, TSR and Signal Aspect

I.	Movement Authority	Suppose XXXX m is available.	XXXX m is to be retained.	If YYYYY m is received and same is to be followed.
II.	SSP/TSR	Suppose Profile X is available.	Profile X is retained	If Profile Y is received, The profile X and Y are required to be superimposed to get a Continuous profile.
III.	Signal Aspect	Aspect is available	Aspect is to be retained on DMI (at least for 6 sec)	New aspect of Signal received shall be shown on DMI

Limited Supervision Time Out:

- In case of retaining the time out of 30 seconds, the firm shall carry out assessment primarily for this activity alone.

6.9 Tags

- Normal Tags only can be linked. Hence, to ensure linking up to the nearest location to the approaching signal, Normal Tags shall be provided in lieu of Signal Approach Tag.
- As far as possible, the adjustment of variation in actual distance to railway specified distance is to be carried out in the Normal tags.
- The maximum distance correction to be incorporated in the Normal tag shall be restricted to 20m.

- If the distance adjustment is required to be done beyond 20m, then adjustment tags shall be provided and Non-Communication Zone shall be introduced to avoid un-necessary emergency braking. The ISA assessment during the application safety case shall deliberate on this aspect and Hazard analysis for the same shall be carried out. This issue is also required to be taken into consideration during pre-installation check.

6.10 Implementation without Line Side Signals

- Three track sections beyond the boundary of the Handing over stationary KAVACH are required to be considered for determination of movement authority.
- The Loco KAVACH shall be able to display the deduced signals based on the track section status. The possibility of display of deduced signals is left to the choice of Zonal Railway.

6.11 Installation

- The Remote Interface Units shall be installed in Double-walled Apparatus Case with proper canopy.
- The Apparatus case shall be painted with coolant paint as circulated by RDSO letter no. STS/E/Apparatus case (fabricated)/Vol II dated 06.06.2012.orlatest
- The developing firm shall furnish the Layout of equipment in Apparatus Case. Redundancy with monitoring shall be provided for Power and Communication channels.
- The Apparatus case shall have provision of lighting and a laptop charging point.

6.12 Criteria for Approval

- Detailed User, Installation and Maintenance manual along with pre- installation, pre-commissioning and Maintenance check lists shall be sub- mitted. Firms shall use these documents during the trials and shall place these documents at the respective stations. The utility of these documents shall be cross checked before the trials are allowed by RDSO. In case of in- sufficiency, the trial would not be permitted. The card wise power supply requirement shall be submitted.
- The RFID layouts, KAVACH control table and wiring diagrams shall be got approved before the start of the trial.
- The dimensions of RIU and proposed Apparatus case along with installation details shall be submitted.

- Since, the equipment is to be provided near track side, all plug-in connectors used, including OFC, shall have a locking arrangement, conforming to suit- able international industrial / automotive / railway specifications or standards. The connectors shall have locks to retain the connection under vibration. They shall be keyed in order to prevent connection in wrong orientation.
- Type testing as per RDSO/SPN/144/2006 shall be carried out for all the equipment including FMS, Communication Modem, Remote Interface Unit and Charger treating this equipment as track side. For battery the highest operating temperature shall be limited to 60°C.

CHAPTER

7 Indigenous Manufacturer

7.1 KAVACH System Consisting of:

(a) Station KAVACH Unit

(b) Loco KAVACH Unit

The loco KAVACH system is located in the locomotive and takes the inputs from the locomotive and controls the locomotive brakes based on the movement authority received from the station KAVACH unit.

Station KAVACH unit is located in a station/IBS or near an LC gate. It takes the inputs like track status, point status etc from the yard and communicates the movement authority to the Loco KAVACH units in its vicinity. It also communicates the gradient data, speed profile till the next approach station.

7.2 The Stationary KAVACH System consists of the following Subsystems

(a) Station Radio Interface unit

(b) Station Master Operation Cum Indication Panel

(c) GPS module

(d) Station Bottom Bins (Optional)

(e) Station Electronic Unit

(f) Remote/LC Gates Interface Unit (RIU)

7.2.1 Stationary KAVACH Equipments

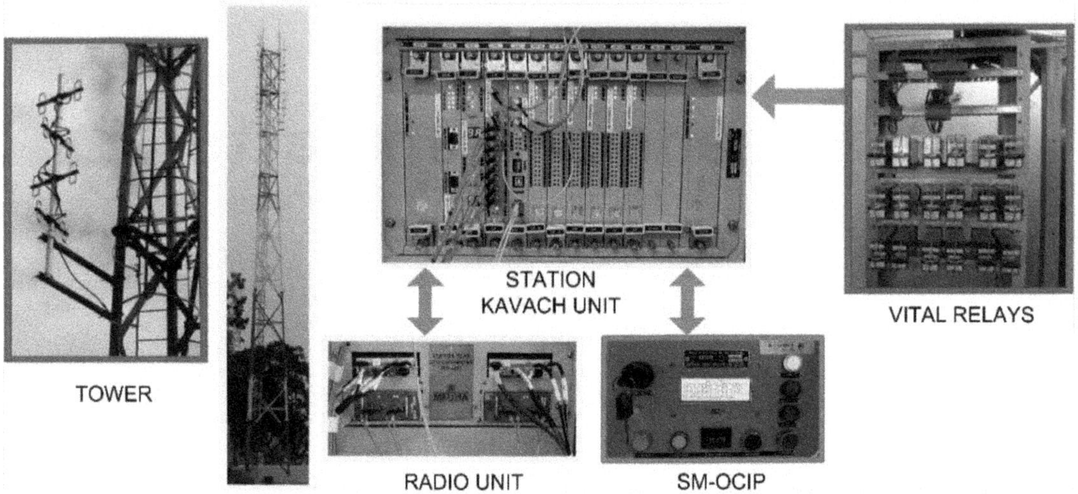

Fig. 7.1

7.2.2 Stationary KAVACH Architecture (MEDHA)

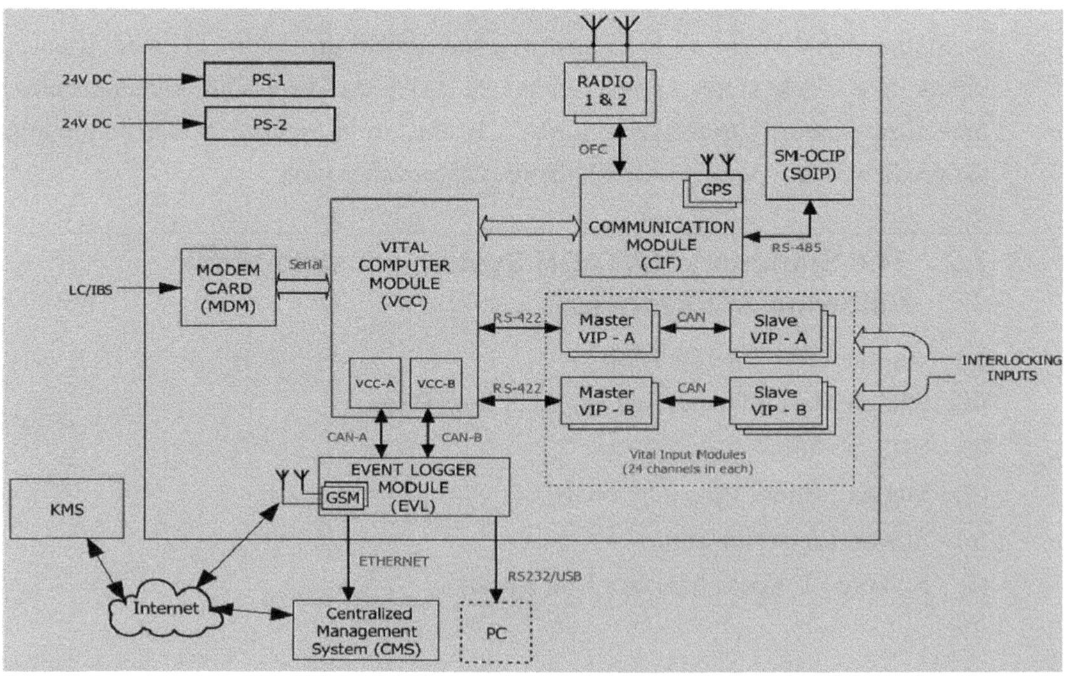

Fig. 7.2

7.2.3 MEDHA – Station Vital Computer

Fig. 7.3

Station KAVACH Electronic unit comprises of following modules

(a) Station Vital Computer (VCC)
(b) Vital Input Card (VIC)
(c) Communication Interface Module (CIF)
(d) Event Logger Module (EVL)
(e) Station Modem Card (SMDM)
(f) Operation & Indication Panel (SOIP)
(g) Power Supply Module (PS)

7.2.4 Stationary KAVACH Module Functions

Vital Computer module - Station KAVACH Vital Computer module is designed as a Dual Electronic Structure based on composite fail-safety with fail-safe comparison. Vital Computer module consists of two vital processors (VP1 & VP2). Two independent computing channels with identical Hardware and common Software perform the safety checking and compare the outputs of two computing channels through inter-processor communication. The outputs are accepted as valid only if the two channels are in agreement.

This module receive the status of signal aspects, position of points, berthing track circuit status and status of Block Instrument Line closed condition through vital input card. Based on the status of field functions, it finds out the current aspect of signals in the yard and calculates the movement authority accordingly. This information will be conveyed to Loco KAVACH unit(s) in the form of Signal information packet.

2oo2 Voting

Two-out-of-two voting (2oo2) also employs two devices. In this arrangement, of the two devices, both devices must "agree" to cause a shutdown before the shutdown will occur. I.e., both devices must vote to trip to cause a trip action. This is physically represented by two switches in parallel. De-energizing either the "A" switch or the "B" switch alone will not cause the entire circuit to de-energize. Only when both switches open is the circuit de-energized and the plant moved to a safe state. This arrangement does not have any tolerance to dangerous failures. A failure of the "A" device in the welded closed mode by itself will result in a dangerous failure of the overall system, and the same is true for the "B" device. While this arrangement does not have any tolerance to dangerous failures, it does have one degree of tolerance to safe failures. If the "A" device were to spuriously fail in the safe open-circuit mode, power will still be conducted through the "B" switch, preventing a spurious shutdown. The same is true for a spurious failure of the "B" switch.

As a result of being tolerance to one safe failure but no dangerous failures, this voting arrangement is commonly used to improve resistance to spurious failures at the cost of decreasing safety performance below what it would be if only a single device were used. As you can see in the table below, use of 2oo2 voting provides for a drastically reduced spurious trip rate, but that improvement in resistance to spurious trips comes with the cost of decreased safety. Specifically, the probability of failure on demand of a 2oo2 voting arrangement is twice as high as for a single device. As a result, this voting arrangement can only be used to reduce spurious trip rates for systems with low SIL requirements e.g., SIL 1.

Vital Input module - Vital Input modules are used for reading the outputs of interlocking system. As 2-out-of-2 architecture is adapted for the design of Stationary KAVACH, each CPU in the 2-out-of-2 is connected to its own set of input cards for reading the field inputs. In each card 24 channels will be present, which can be used for reading NO contacts of field relays. Number of input cards for each system may vary from 1 to 32 cards.

The input contacts are isolated from the CPU data lines by using opto-isolators.

Each vital input card is capable of reading up to 24 inputs and Employed double cutting mechanism. Fail safe input read-back circuit, Inbuilt Surge Current protection, reverse Polarity protection are provided.

Communication Interface module - Communication module is used for performing communication between the Vital Computer module and other sub-systems.

Communication module has serial communication interface with Radio modems, GPS modules and SOIP communication

Event Logger module - The Event logger Card will be used as a Data recorder. This module communicates with Vital Computer module to get the system events. It will record system events and the fault information of all control modules in the Station KAVACH unit. The Event logger Card will communicate with Centralized Management System through GSM and Ethernet interface.

Modem Card - Modem card provides Quad cable interface between the two Station KAVACH units. Communication is a full-duplex with a single pair of quad cable.

Modem card provides Serial interface between the Station KAVACH and IB/LC/Hut (RIU) units. Communication shall be a full duplex with a single pair of OFC fiber. Modem card shall receive serial data from KAVACH Vital computer card, received data shall co- transmit it to VIPs placed at IB/LC/Hut (RIU) units. Data received from VIPs at IB/LC/Hut units shall send it to station KAVACH Vital computer module.

Power Supply module for Top Bin - The Power Supply module will generate the voltages required for the operation of all the modules in the system from +24V DC Input Supply. This module will work with a voltage range of +24V DC (+30%, -20%).

Station KAVACH Top Back- plane - Backplane provides the necessary interface between all the cards in the system and connects the system to the other systems.

Power Supply module for Bottom Bin - The Power Supply module will generate the voltages required for the operation of all the modules in the system from +24V DC Input Supply. This module will work with a voltage range of +24V DC (+30%, -20%).

Station KAVACH Bottom Back-plane - Backplane provides the necessary interface between all the cards in the system and connects the system to the other systems.

Redundancy: Redundancy plays a key role in increasing the availability of a system, without compromising the safety aspects of the system.

Stationary KAVACH Radio Interface Units

The Stationary KAVACH comprises of Radio Interface units, located near to the tower. Radio Interface units consisting of two Radio modems and two OFC to RS-232 converters. The Radio modems used have two ports- one for communication and the other for diagnostic port to detect the health status. The Vital processors have interfaces to these ports.

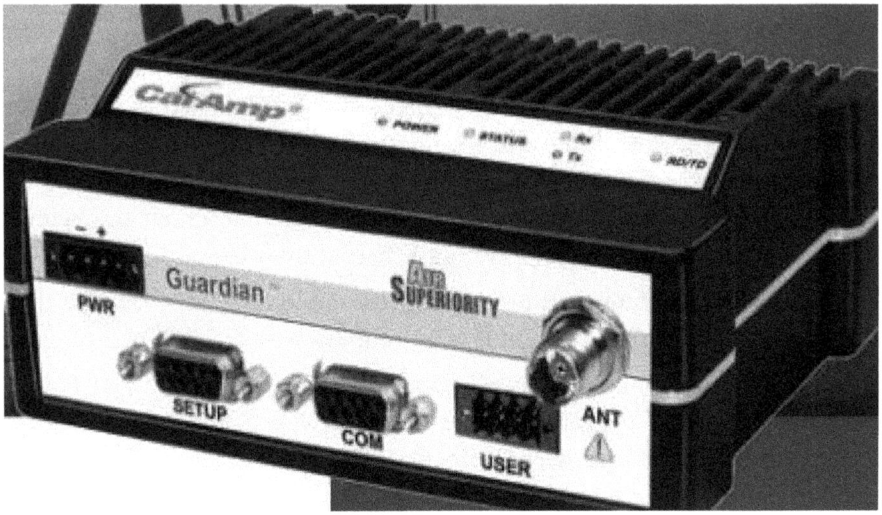

Fig. 7.4 Radio Modem.

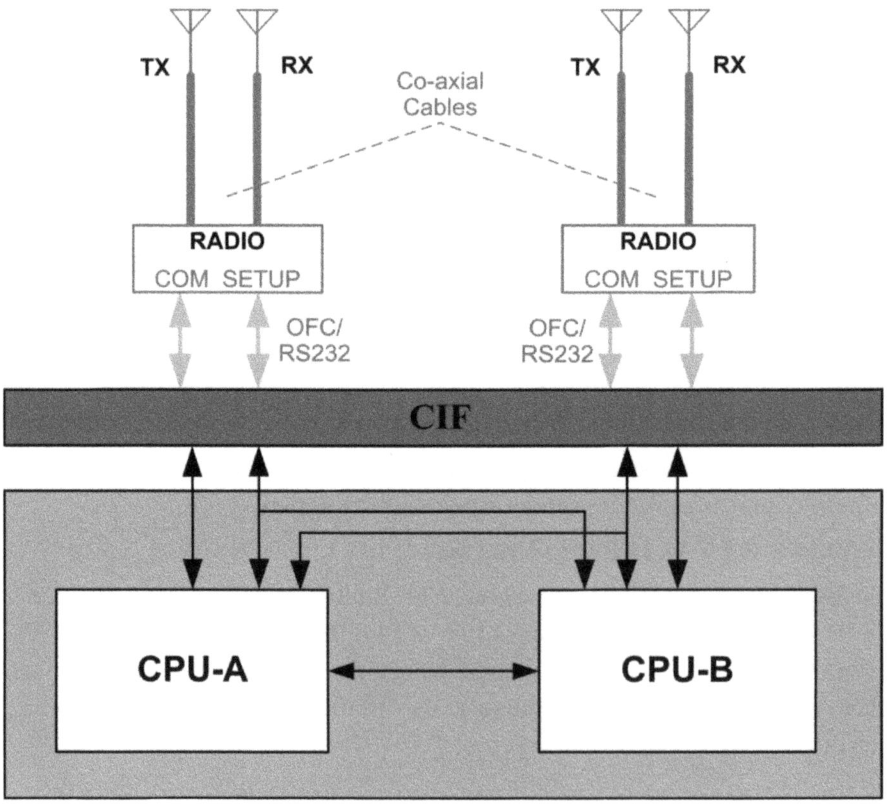

Fig. 7.5 Radio modem Interface

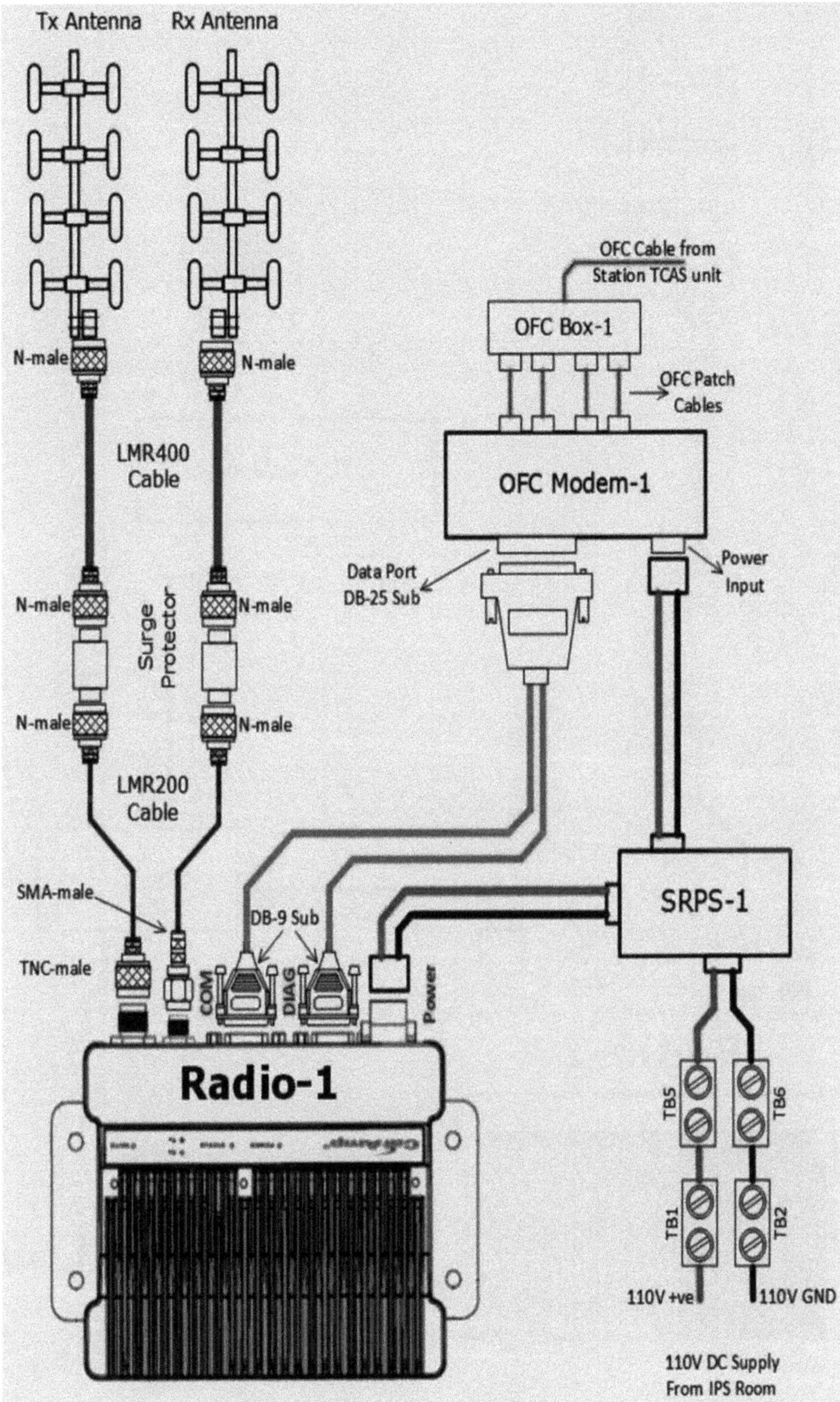

Fig. 7.6 Antenna wiring from Radio1 in Location Box

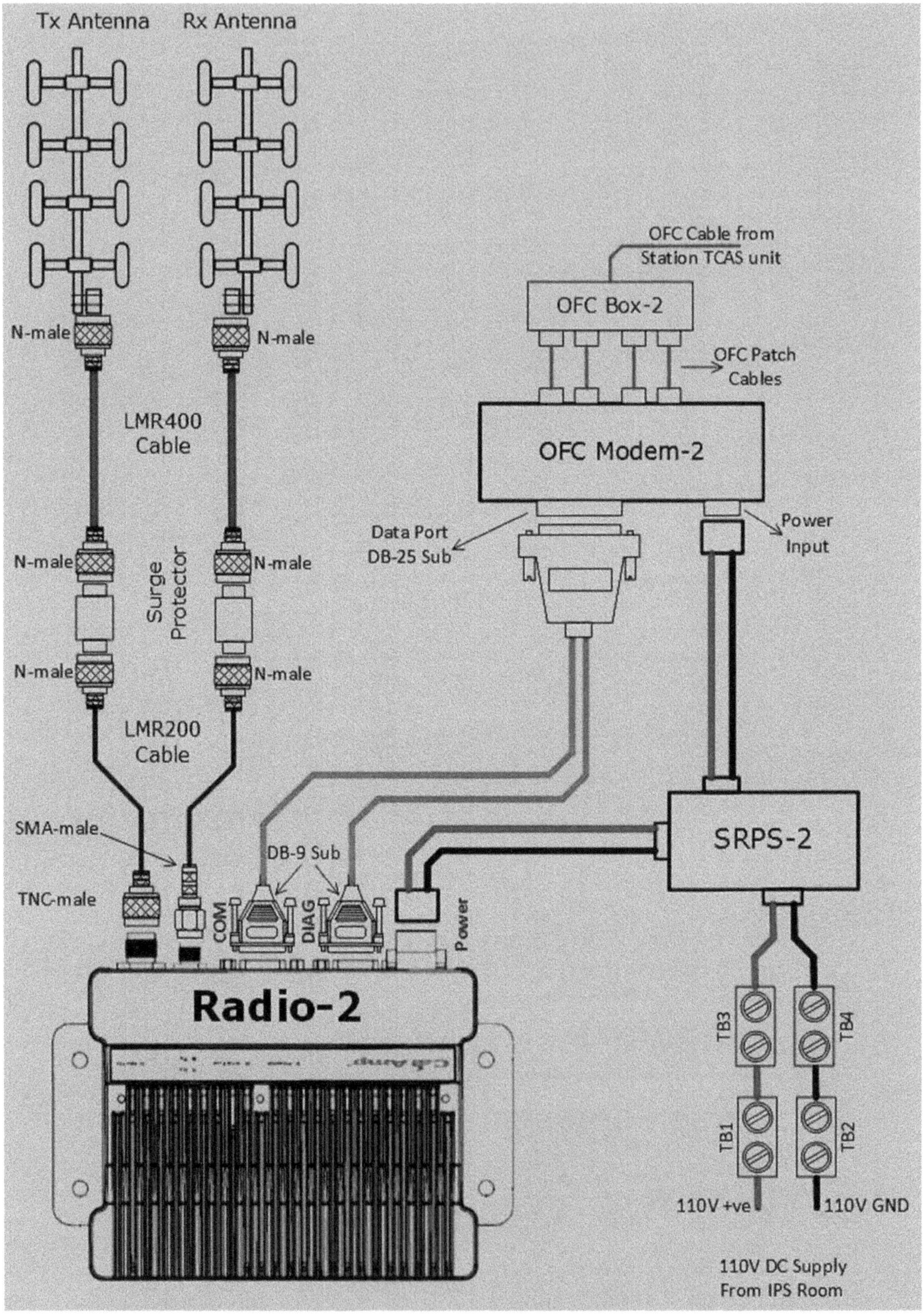

Fig. 7.7 Antenna wiring from Radio2 in Location Box

Station Radio Interface Unit - Station radio forms the important part of the Station KAVACH unit. Station radio is used for communication between the Loco KAVACH and Stationary KAVACH units. Station KAVACH unit is connected to the Radios by using the Communication Interface Module (CIF) and Station Radio Communication Interface Unit (SRCU) The radio interfaces with two dipole antennas – one for transmit and the other to receive the data. The two antennas are mounted on a tower which stands at a height of 30 meters to 40 meters. The Radio modems and SRCU modules are housed in a Tower box which is mounted on the antenna tower at a few feet height above the ground. The antenna cables coming from the Radio antenna and the OFC cables from the CIF module are terminated at the Tower box. The CIF module of the Stationary KAVACH unit is placed in the relay room. The CIF interfaces with the SRCU by using four Multi Mode OFC cables – 2 for data, other 2 for diagnostics. The OFC cables are terminated at the location box and the optical signals are converted to electric signals by using the SRCU module. The SRCU has two sets of RS-232 interfaces (one for Loco communication data and the other for Radio diagnostics). The RS-232 signals are connected to the Communication port and set-up port of the Radio modem there is another similar setup for interface with the second radio modem, with the antenna tower and the CIF module being common for both the setups

Station Master Operation and Indication Panel (SM-OCIP)

The SMOCIP is an operator (Station Master) interface module placed within the reach of the Station Master. The interfaces that are provided are an LCD, a Key actuator, three Push buttons, six digit non-resettable counter and LED lamp indications.

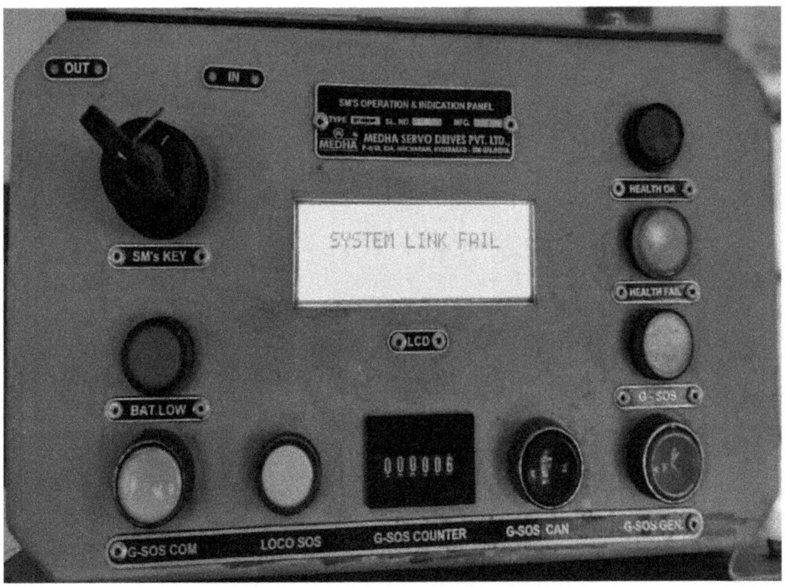

Fig. 7.8 SM-OCIP Front View

The Push buttons of the SMOCIP module are powered only when the SM key is turned to IN position. The Station Master is authorized to give commands like SOS generation and SOS cancellation using these three push buttons labeled as SOS Gen, SOS cancel and SOS common. Commands are accepted by the Stationary KAVACH only if the SOS common button is pressed in conjunction with SOS Gen/Cancellation button. This is to prevent form the accidental generation of SOS message.

The six digit non-resettable is used to register the count of SOS generation and cancellation operations. LED lamp indications are used to show the system health, battery health and SOS status are present in SMOCIP.

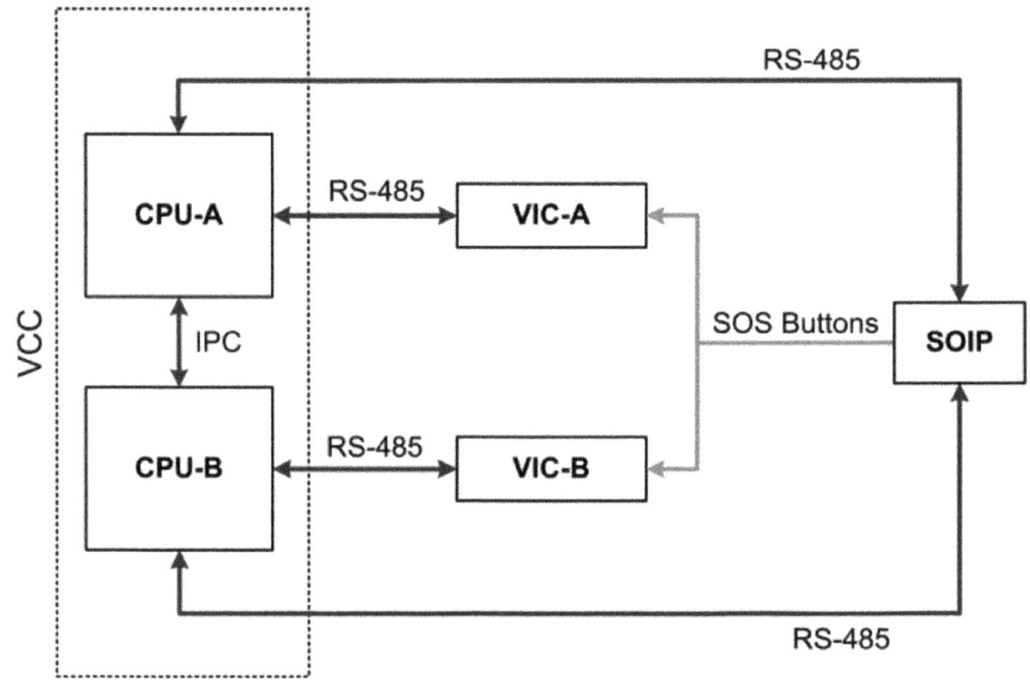

Fig. 7.9 SOIP Interface Diagram

7.2.5 SMOCIP module performs the following functions

1. Displays system information on LCD
2. Used by station master to generate and cancel the station general SOS using the push buttons
3. Displays the loco SOS information on LCD
4. Alerts the station master with a buzzer, when events like channel link fail, button stuck, critical faults, general SOS or loco SOS occurs
5. Records the station general SOS operation using counter
6. Indicates system health and SOS active/ inactive status using LEDs.

LED's Indication

1. HEALTH OK LED Green colour indicates communication health status. If the communication is healthy between SMOCIP and SVCC, this LED turns ON, otherwise it remains OFF.
2. HEALTH FAIL LED Red colour indicates communication failure status. If there is communication failure between SMOCIP and SVCC, this LED turns ON, otherwise it remains OFF.
3. G-SOS LED Red colour indicates General SOS status. If General SOS is active, this LED turns ON, otherwise it remains OFF.
4. LOCO SOS LED Yellow colour indicates loco specific SOS status. If Loco specific SOS is active, this LED turns ON, otherwise it remains OFF.

Push Buttons

1. Three push buttons are operable only when SM key is inserted and is in 'IN' position.
2. G-SOS COM push button is used for general SOS generation and cancellation. It is also used as acknowledge button to acknowledge the events.
3. G-SOS CAN push button is used for general SOS cancellation.
4. G-SOS GEN push button is used for general SOS generation.

SMOCIP displays the following information on LCD screen, based on the detected event or information received from SVCC.

1. Checksum Display
2. Button Stuck Display
3. General SOS Display
4. Loco SOS Display
5. Critical Faults Display
6. Faults Display
7. Recovered Faults Display
8. Channel Link fail Display
9. System Link fail Display

7.2.6 Station Antenna

Base station stacked Dipole type Omni directional type antenna with Gain of 9.0 dB. Frequency band is 406-512 MHZ with input impedance 50 Ohms. Four antennae used and 1- TX and 1- RX antenna for each modem.

GPS & GSM Module: GPS is used for time reference of Station KAVACH. It is housed in the Communication module, which communicates with the vital computer to give GPS related information. The GPS receiver outputs the data which is routed to the Vital Computer Card after undergoing proper isolation in the CIF card. GSM is used to communicate with KMS server to get the Authentication keys. It is housed in the Event Logger module, which communicates with the KMS server.

GPS & GSM Antenna: The GPS chip set receives the GPS data with the help of GPS antenna which captures the information transmitted by the GPS satellites. The GPS antenna used is a patch antenna which is terminated with an SMA male connector. This GPS antenna is connected to the female SMA connector mounted on the CIF card whose other end is connected to the Pin 16 (Antenna Input RF IN) of the GPS chip set. To get the GPS information from maximum number of satellites, it has to be placed in a location where the GPS antenna has maximum possible sky view. The GPS & GSM antenna is enclosed in an Antenna Hood which is usually mounted on the top of the room containing Station KAVACH cabinet.

GPS INTERFACE: GPS interface used for Time synchronization of KAVACH Radio communication network. Two GPS Modems are used for Redundancy and both are active. Faulty one detected and isolated with in 3 cycles and auto recovery of faultyGPS. Real time events logging also possible.

GSM: GSM used for Events & Faults loggings in NMS through GPS and to get security keys from KMS, to send SMS for emergency

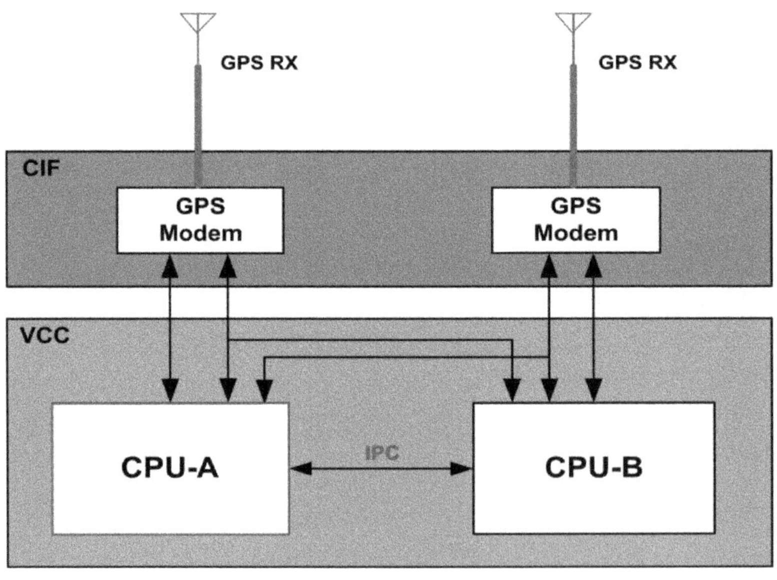

Fig. 7.10 GPS Interface

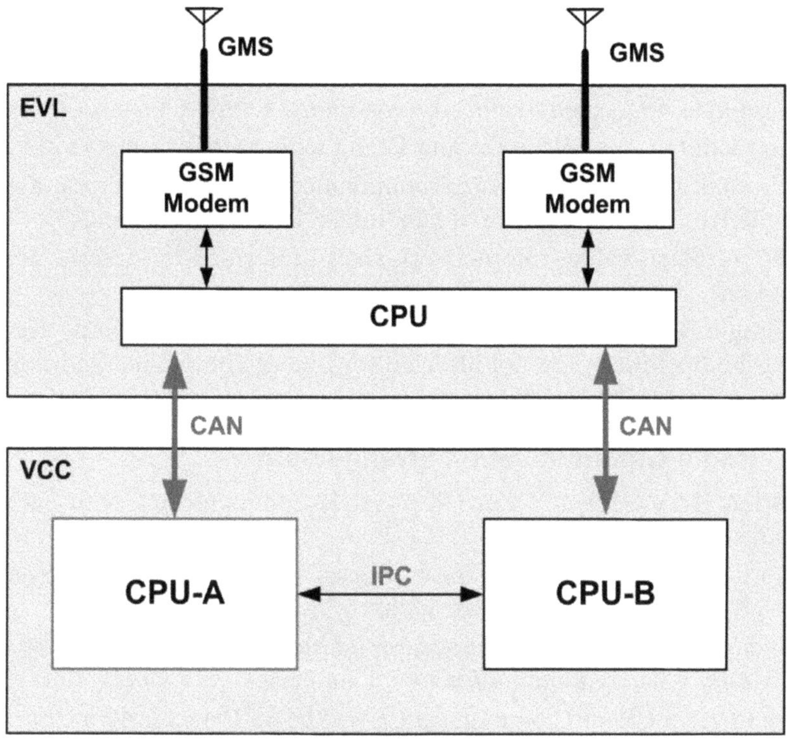

Fig. 7.11 GSM Interface

1. Two GSM modems are used for redundancy
2. Events and Faults logging in NMS through GSM
3. To get Security Keys from KMS
4. To send SMS for emergency/critical messages

Fig. 7.12 Placement of GPS & GSM Antennas

7.2.7 Remote Interface Unit (RIU)

- Remote Interface Unit (RIU) shall be used where remote signalling functions are required to be fetched to a nearby Stationary KAVACH unit.
- RIU shall have provision for interfacing with signalling inputs in fail-safe manner.
- RIU shall use line modems for communicating with stationary KAVACH unit. A single RIU shall be capable of handling at least 32 field inputs
- OFC or Quad cable, ring network shall be formed to increase availability of the network.
- A single RIU shall be capable of communicating with two adjacent RIU units so that the operations are not affected in case of communication link failure on one side only. It is 2oo2 architecture

7.2.8 Radio Communication arrangement

- Station KAVACH and Loco KAVACH communicates with each other through Radio modems.
- The Radio with a transmission power of 10W is capable of communicating with the other KAVACH within the radius of 5 KM atleast.
- Antenna for station/IBS/mid-section interlocked Gate unit will be Combination of vertically polarized Omni-directional antennae.
- The tower for Radio communication will be of 30 meter/40 meter height
- Station KAVACH and Loco KAVACH units communicates with each other for every 2 seconds.
- Pair 1 : Station Tx - 456.8 MHz, Rx - 416.8 MHz
- Pair 2 : Station Tx - 466.8 MHz, Rx - 426.8 MHz
- Common Frequency, f0 - 441.8 MHz

7.2.9 Remot Interface Unit (RIU)

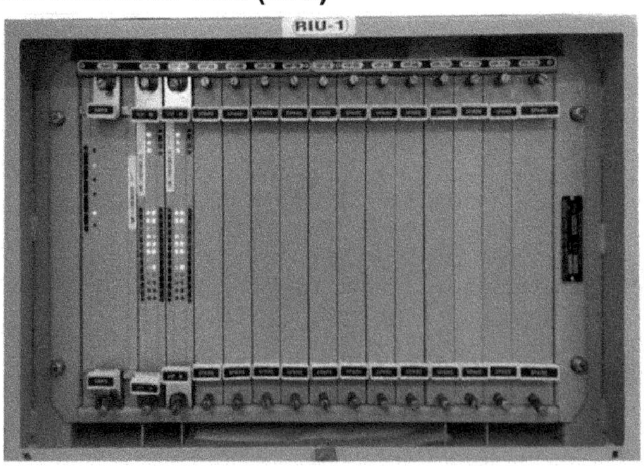

Fig. 7.13

Remote Interface Unit (RIU) comprises of the following sub systems

1. Equipment Rack
2. RS485OFBC Converter (M485OFBS)
3. Battery Charger
4. Battery Set

Each RIU consists of four Vital Input Modules (VIP) and each VIP module is reads maximum of 24 inputs. There- fore, station KAVACH reads total 96 field inputs from each connected RIU subsystem. Remote signaling inputs are communicated to Station KAVACH unit over an OFC media. If multiple RIUs are required to communicate with Station KAVACH unit, they are to be connected in a ring network topology to increase the availability of the network. In ring network, each RIU is connected to two adjacent RIUs. In case, communication link fails with one adjacent RIU, inputs are communicated over the other communication link. The same set-up is used to interface LC gate inputs to stationary KAVACH unit. Equipment rack is used to fetch the field inputs of relevant IB HUT/LC gate. All the modules are plug-in type and 6U high that plugs into a Back Plane (SBBP) in the rack.

RS485 to OFC Converter (M485OFBS)

RIU bins consist of full duplex RS485 communication interface to communicate with the Vital Computer Card of S KAVACH. This module consists of two RS485 to OFC conversion sections - Primary and Secondary. This module facilitates the connection of RIUs in ring network topology. RIU consists of the following modules

S. No.	Module Name	Functions
01	Vital Input module (M110VIP)	Vital input module reads the outputs of interlocking system. Each card contains 24 channels, which reads NO contacts of field relays. The set of input modules consists of an interface to signal aspects, berthing track circuit, and LC gate status condition through potential free contacts. Each vital input module can read maximum of 24 input relays and indicates the status of all inputs. All the input contacts are isolated from the field input lines using opto-isolation of 2 kV. Vital input module scans and de-bounces the input relay status and form the input message to send the same to vital computer module of Station KAVACH. Vital input module communicates the status of inputs in every cycle over ring network.
02	Power Supply module (M110SBPS)	The power supply module generates the voltages required for the operation of all the modules in the system from +24 V DC input supply. This card provides isolated output power of 4.5 V ± 5% to VIP cards in the system. This module works with an input voltage range of +24 V DC (+30%, -20%).
03	Bottom Back Plane (M110S BBP)	Backplane interfaces with all VIP cards in the system and connects the system to the other.
04	Field Interface Card (M110FIC)	FIC card interfaces with field inputs to VIP card through FRC cable. Each FIC card is capable of interfacing 48 inputs to RIU unit for CH-A and CH-B.

7.2.10 Interfacing Input relays

Inputs are read by RIU unit through field interface card. The cable 50 wire jacketed flat ribbon cable of length one metre connects FIC card to 50 pin shrouded header K19 on the station KAVACH RIU unit back plane. The field inputs from relays are given to FIC card using CABLE SET 14 (bunch of cables) to the terminal blocks on the FIC card. If more number of inputs are to be connected, an arrangement similar to cable Set 14, cable 12 are used to connect the inputs to RIU back plane at connector K22.

7.2.11 Interlocking (IXL) System Inputs Interface

- Only Signal Aspects, Berthing Tracks, Point Machines
- Each vital input card is capable of reading upto 24 inputs
- Capable of reading 784 inputs (32 VIP Cards)
- Employed double cutting mechanism
- Two independent channels (2oo2 architecture)
- Fail safe input read-back circuit
- Inbuilt Surge Current protection, Reverse Polarity protection

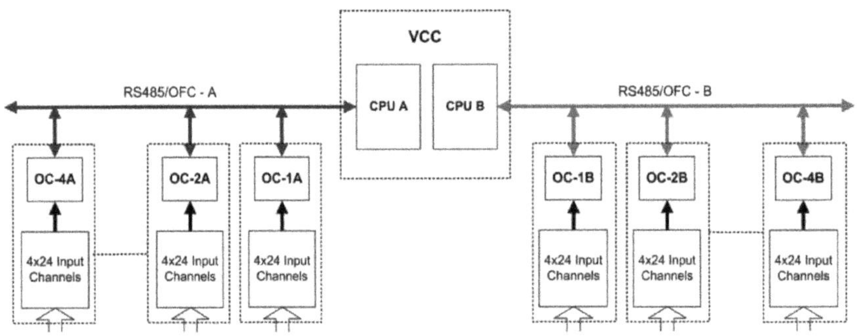

Fig. 7.14 Interlocking system inputs interface

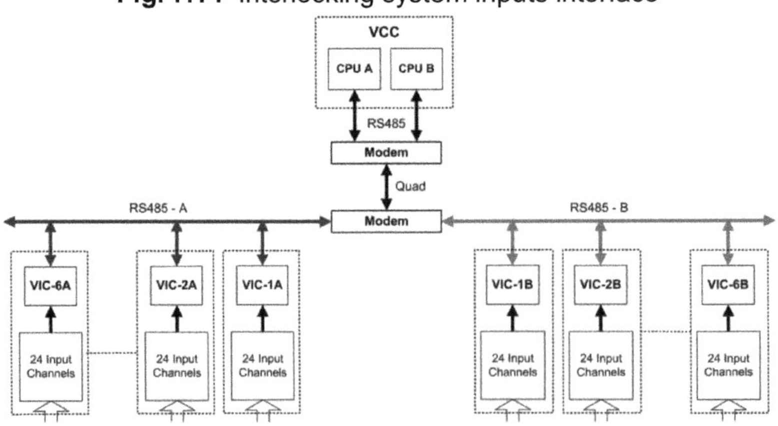

Fig. 7.15 LC Gate/ IB hut input interface

7.3 Loco KAVACH Unit

The loco KAVACH system is located in the locomotive and takes the inputs from the locomotive and controls the locomotive brakes based on the movement authority received from the station KAVACH unit.

The Loco KAVACH unit comprises of the following sub modules.

1. Loco KAVACH Electronic Unit
2. Loco KAVACH Radio Interface Units
3. Speed Sensor
4. RFID Readers
5. Driver Machine Interface (DMI) Units
6. The Loco KAVACH Electronic unit
7. The following table lists the modules housed in Loco KAVACH Electronic Bins and their basic functionalities.

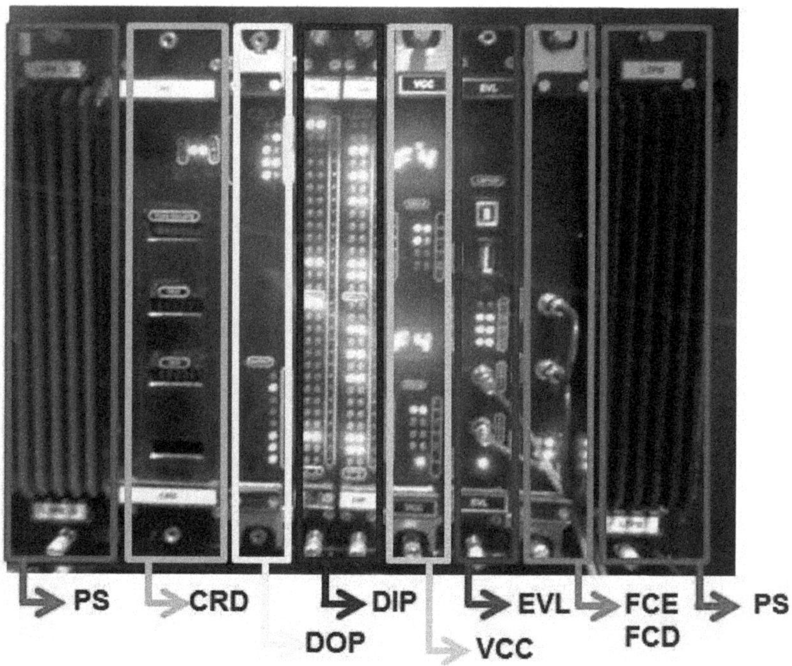

Fig. 7.16 LocoKAVACH Control Unit (MEDHA)

S.No.	Module Name	Basic Function(s)
1	Vital Computer module (M110VCC)	KAVACH consists of One Vital Computer module, with 2 Vital Processors running in 2-o-o-2 architecture with diverse hardware. System Inputs are read and processed independently from respective Input Channels of the Input Cards to give functional Outputs.

Contd...

S. No.	Module Name	Basic Function(s)
2	Digital Input module (M110DIP)	KAVACH consists of two input cards. Each input card has 28 channels, with 14 channels corresponding to each vital processor. 14 channels in an input card are connected to the same input as the 14 other channels of the input card. The input contacts are isolated from the CPU data lines by using opto-isolators.
3	Digital Output module (M110DOP)	The Digital Output module is used to give different output of the system. The outputs are used to drive emergency relay, interface with the Brake Interface Unit (BIU) and perform non-vital operations like driving counters, buzzers, status indicating lamps. The Digital Output module receives the vial output drive data from the Vital Computer module.
4	Frequency Input and Communication Interface module (M110FCD for Diesel loco and M110FCE for Electric loco)	The FCD/FCE card is used to interface the VCC with external systems using, by converting the TTL level serial communication signals to RS485 signals, isolated from the VCC. This card consists of the GPS section, which the controller uses for time reference. This card consists of frequency input sections, which receive the frequency signals from the speed sensor and convert them to TTL level. This card also houses the power manager section which cuts off the supply to the controller in case of controller failure. Power manager also performs the important task of power sequencing.
5	Counter and Relay Driver module (M110CRD)	The Counter and Relay Driver module consists of the counters and emergency relay related to the system. Counters are used to maintain the count of different parameters of the system.
6	Event Logger module (M110EVL)	Event Logger module will be used to store the system events. This module is also used to communicate to the KMS and NMS using GSM. This module is interfaced to both the Vital processors through serial communication (CAN).
7	Loco KAVACH Power Supply module (M110LTPS)	The system consists of two redundant Power Supply modules. The nominal input voltage to the card is +72V dc (Diesel Loco) and +110V dc (Electric Loco). These cards provide isolated power to all the cards in the system. Over voltage and Under voltage protections for the input and Over voltage and Overload protections for the output are provided.
8	Loco KAVACH Back-plane (M110LCBP)	Backplane provides the necessary interface between all the cards in the system and connects the system to the other systems.

7.3.1 Loco KAVACH Radio Interface Unit

The Loco KAVACH comprises of two Radio Interface units with each unit consisting of one Radio modem, one RS-422 to RS-232 converter and the associated power supply. The Radio modems used have two ports- one for communication and the other for diagnostic port to detect the health status. The Vital processors have interfaces to these ports. The Vital Processors can configure the Transmit and Receive frequencies, request and receive health information from the Radio modems.

7.3.2 RFID Readers

The Loco KAVACH unit interfaces with the two RFID readers to read the Data programmed in the RFID tags which are mounted on the Tracks. The RFID tag is a passive device which is activated by the RFID reader. There are two RFID readers interfaced to the Loco KAVACH. The Vital Processors configure the RFID reader to read two pages of width 64-bits each at one instance. The communication protocol between the Vital Processor and the RFID reader is as per the serial protocol defined in RFID Reader manual.

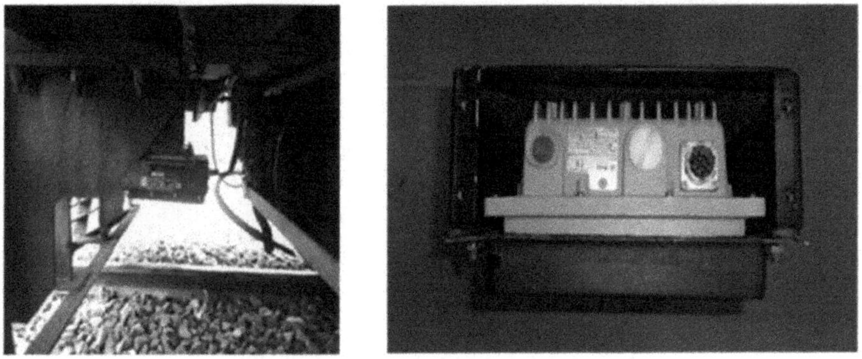

Fig. 7.17 RFID Reader

Fig. 7.18 LOCO – Radio Antenna

1. Gain 0 dB, Frequency band : 406 – 512 Mhz
2. Input Impedance: 50 Ohm
3. Polarization: Vertical
4. Radiation pattern: Omni directional

5. Wind rating: 200 km/hr
6. Weight (Kg): 3
7. 4 Antennae (1-TX and 1-RX antenna for each radio modem)

7.3.3 Speed Sensor Interface

The Speed Sensor is mounted on the axle of the locomotives. Based on the wheel diameter (configurable) and number of rotations per second, the speed will be calculated. The speed sensor also has the ability to give the direction of the locomotive. The speed sensor has redundancy built into it.

Fig. 7.19 Loco Speed sensor unit

7.3.4 Power Supply Module

Power Supply module converts 72/110V dc supply to the supplies required by the sub-systems of the Loco KAVACH unit. Nominal working voltage is 72V (for Diesel) & 110V (for Electric). Input working voltage range 20V to 100V (for Diesel) 60V to 160V (for Electric)

7.3.5 MEDHA-RFID

Fig. 7.20

Fig. 7.21

7.3.6 Loco to Station Radio Communication Requirement

Loco KAVACH unit shall update its position to Station KAVACH unit within station vicinity. Station vicinity is defined as 3.5km from Home Signal or Two Normal Tags before the Distant signal. Loco to Station communication is mandatory in this vicinity.

7.4 Station KAVACH System Indications

Before set up and configuration of MCA110 system, we need to know front panel controls and indications associated with MCA110 system printed circuit boards. The front facia of all the cards are provided with LED indications. These indications assist the operator to know the status of the card at any point of time. The following tables provide detailed information about the indications and other components provided on the facia of station KAVACH unit cards.

Station KAVACH vital computer card indications

Table 7.1 Vital input card indications

Indication/ Interface	Description
LED(POWER)	Indicates Power(3.3V) input to the card
LED(CYC)	Indicates Cyclic activity status, toggles in every cycle
LED(IPC)	Indicates Communication status with other CPU
LED(GPS-1)	Indicates Communication status with GPS-1
LED(GPS-2)	Indicates Communication status with GPS-2
LED(EVL)	Indicates Communication status with EVL
LED(VIP)	Indicates Communication status with VIP Cards
LED(OCIP)	Indicates Communication status with OCIP
LED(RAD)	Indicates Communication status with Radio
7 Segment Display	Displays Fault codes

Table 7.2 CIF Card controls and indications

Indication/Interface	Description
LED(POWER)	Indicates Power(3.3V) input to the card
LED(CYC)	Indicates Cyclic activity status, toggles in every cycle
LED(COM)	Indicates Communication status with vital computer card
LED(IP-1 to IP-24)	Indicates Input-1 to Input-24 status (Pick- up/Drop)

Table 7.3

Indication/Interface	Description
Connector (GPS-B)	Used to connect the Antenna for GPS Receiver – 2
Connector (GPS-A)	Used to connect the Antenna for GPS Receiver – 1
OFC Connector (VCB-2RX)	Used to connect the Radio-2 Diagnostic port receiver
OFC Connector (VCB-2TX)	Used to connect the Radio-2 Diagnostic port Transmitter
OFC Connector (VCB-1RX)	Used to connect the Radio-2 COMP port receiver
OFC Connector (VCB-1TX)	Used to connect the Radio-2 COMP port Transmitter
OFC Connector (VCA-2RX)	Used to connect the Radio-1 Diagnostic port receiver
OFC Connector (VCA-2TX)	Used to connect the Radio-1 Diagnostic port Transmitter
OFC Connector (VCA-1RX)	Used to connect the Radio-1 COMP port receiver
OFC Connector (VCA-1TX)	Used to connect the Radio-1 COMP port Transmitter
LED(VHMPWR)	Indicates Voltage Monitoring Circuit Power
LED (INTPWR)	Indicates Power (3.3V) input to the card
LED (V STS)	Indicates Vital Computer Card Voltage Monitoring Status
LED (V STS)	Indicates Vital Computer Health Status

EVL CARD Indications

Table 7.4

Indication/Interface	Description
LED(POWER)	Indicates Power (3.3V) input to the card
Connector (GSM-1)	Used to connect the GSM-1 Antenna
Connector (GSM-2)	Used to connect the GSM-2 Antenna
LED(CYC)	Indicates Cyclic activity status, toggles in every cycle
LED(ETH)	Indicates Communication status with E1 interface
LED(VCA COM)	Indicates Communication status with Vita computer-A
LED(VCB COMP)	Indicates Communication status with Vita computer-B
LED(GSM-1)	Indicates Communication status with GSM -1
LED(GSM-2)	Indicates Communication status with GSM -2
LED(PENDRV)	Indicates Pendirve connectivity status
LED(LAPTOP)	Indicates PC/Laptop connectivity status
USB Connector (PENDRIVE)	Used to connect Pendrive
USB Connector (LAPTOP)	Used to connect PC/Laptop
Ethernet Connector	Used to connect adjacent station KAVACH through E1 interface

MDM Indications

Table 7.5

Indication/Interface	Description
LED (POWER)	Indicates Power(3.3V) input to the card
LED (STN-1 to STN-6)	Indicates Communication status with Adjacent Station/IBS/LC gate
Connector (DB-9)	Used to connect Quad-cable (2-wire)
Connector (DB-9)	Used to connect Quad-cable (2-wire)

Battery Charger card Indications

Table 7.6

Indication/Interface	Description
FAIL	Battery Charger output fail status
O/P	Battery Charger 24V output status
I/P	Battery Charger 110/230V AC Input status
LOAD	Battery Charging status
V.ADJ	Battery Charging Voltage Adjustment POT

7.5 System Start-up Details

After configuring and installing MCA110, the next step in the start-up process i.e. applying power to the system, and then verify that the internal functions of the MCA110 system is working properly.

Initial Checks

The following checks should be made before switching ON the system:

(a) No loose wires are hanging from any of the PCBs.

(b) Indications on the facias of all the PCBs are in OFF condition

(c) No PCB is projecting outside. The fixing screws for all the PCBs are fixed properly

(d) Field power cable is connected to PS Card and required Relay contacts.

(e) Relays are inserted in their respective slot.

(f) User Interface unit is to be connected properly.

(g) Communication and Power cables are properly terminated.

7.6 System Maintenance

This chapter provides instructions for performing a scheduled maintenance routine that is designed to detect early symptoms of equipment degradation.

Maintenance personnel must be familiar with the recommended methods for operating, testing, and repairing the MCA110 system equipment.

7.6.1 Tools

Following Tool kit shall be purchased separately.

Table 7.7

S. No.	Tool	Quantity
1	Digital multi-meter	1 No.
2	Ring spanner	1 No.
3	Open end spanner	1 No.
4	Socket spanner with handle	1 No.
5	Screw Driver No. 902	1 No.
6	Screw Driver No. 935	1 No.

7.7 Periodic Maintenance

- Service Engineer shall download Events and Faults log and store in a PC with date and time.
- Wiring shall be verified to avoid loose contacts.
- Nuts and bolts shall be verified for tightness at cabin.

- Input Supply shall be monitored on the Terminal board and it shall be with in specified limits.

7.7.1 Electrostatic Protection

- Always stand on an approved conductive floor mat when touching or handling printed circuit boards.
- Always wear a wrist strap-grounding device. The wrist strap should have a 1.0MΩ current limiting resistor. Connect the wrist strap grounding connector to suitable ground connection.
- Periodically check each wrist strap for continuity using an approved tester. Continuity readings must be between 500kΩ and 10MΩ. Discard any wrist strap that does not meet this criterion.
- Always handle printed circuit boards by the edges. Do not touch board components.
- Once removed from the bin, immediately place printed circuit boards into a conductive-shielded bag. Wrap the bag in conductive foam to protect the circuit board during transport and shipment.

7.8 System Trouble Shooting

Troubleshooting the MCA110 system involves careful analysis of observable fault codes displayed on Station SM's OCIP LCD, and the ability to perform standard electrical and electronic troubleshooting operations.

System event log information associated with different modules are captured in Application terminal tool program contain information regarding the performance of MCA110. These event logs are useful while troubleshooting. The information placed in the event logs contain Fault Codes, which contain both Critical and Non Critical Faults.

The Critical Faults are the Faults, which may cause Restart or Shutdown of the Faulty module, or they indicate a serious error, which needs immediate attention.

The Maintainer is alerted about the Critical Fault by a Fault Buzzer in the OCIP Non Critical Faults are the faults, which do not cause the System to restart, or Shut-down, but the System may continue to operate in a de-graded mode of operation.

Trouble shooting each module

This section provides the troubleshooting actions to be taken for the fault codes observed on each Module.

Trouble shooting Vital Computer Card (M110VCC-01) faults

Table 7.8

Fault code on LCD/ 7 Segment display of MCC	Fault Code description	Trouble shooting action
X: FC_PBIST(0x01) Where X can be CPU A/B Note: Fault string displays in LCD and Hex code displays on 7 segment display of VCCs	Power on built in self test fail	Replace the VCC card
X:FC_DIP_SW_FLT(0x02)	Dip switch fail fault	
X: FC_CAN1_LPBK (0x03)	CAN1 loop back test fail	
X: FC_CAN2_LPBK(0x04)	CAN2 loop back test fail	
X:FC_DMA_MODULE (0x06)	DMA module test fail	
X: FC_WDT (0x07)	Watchdog test fail	
X: FC_CYC_TIMER(0x08)	Cycle timer test fail	
X: FC_POST_TIMER(0x09)	Post timer test fail	Replace the GPS Antenna
X: FC_CRC_MODULE(0x0A)	CRC module fail	Replace the VCC card persists replace the CIF card. Still problem persists replace the VCC Card
X: FC_FPGA_PWRUP(0x0B)	FPGA power up fail	
X: FC_SLF_TST_TMOUT(0x28)	Self test timeout	Replace the VCC card

Trouble shooting Vital Input Card (M110VIP-01) faults

Table 7.9

Fault code on LCD	Fault code description	Trouble shooting action
FC_DIP_SW_FLT Note: Fault string displays in LCD	Dip switch configuration fail	Replace the VIP card
FC_CAN1_LPBK	CAN1 loop back test fail	
FC_INT_FLASH	Internal flash test fail	
FC_INT_RAM	Internal RAM test fail	
FC_WDT	Watchdog timer test fail	
FC_CYC_TIMER	Cycle timer test fail	
FC_POST_TIMER	Post timer test fail	
FC_CRC_MODULE	CRC module test fail	
FC_SCI_LIN_LPBK	SCI LIN loop back test fail	
FC_SLF_TST_TMOUT	Self test timeout	

Trouble shooting Power Supply Card (M110STPS-01) indications

Front facia LEDs	Trouble shooting action
LED(POWER)	If Power LED is blank Press faulty PS card tightly. If problem persists, check Filter output voltage for 24V, if there is no output voltage, change the faulty filter.
LED(ERR)	If this LED is glowing Red Press faulty PS card tightly. If problem persists, check Filter output voltage for 24V, if it is not correct, change the faulty filter else change the faulty Power supply card.
LED(4.5V)	If 4.5V LED is OFF Press faulty PS card tightly. If problem persists, check Power supply card 4.5V output voltage, if it is not correct, change the faulty Power supply card(PS)
LED(6.2V)	If 6.2V LED is OFF Press faulty PS card tightly. If problem persists, check Power supply card 6.2V output voltage, if it is not correct, change the faulty Power supply card(PS)
LED(ISO 5V)	If 5V LED is OFF Press faulty PS card tightly. If problem persists, check Power supply card 5V output voltage, if it is not correct, change the faulty Power supply card(PS)

Note: please refer trouble shooting manual for further details

Do's and Don'ts

S. No.	Do's
1	Always refer correct documentation.
2	Always use recommended tools.
3	Log sheet shall be regularly updated and kept as future reference.
4	The steady battery voltage at 24 V should be maintained for Station KAVACH
5	Ensure that all cards mounting screws are tightened properly
6	Under regular maintenance Download the events from EVL for every one month
7	Use the ESD footwear in case of Cards replacement or repair

Contd...

S. No.	Don'ts
1	Don't remove the modules from Station KAVACH System is under Power ON condition. Remove the modules if necessary after switching OFF the power to system.
2	Repair of cards is a highly technical job and is not possible at site. Hence Railways should not carry it out. The card should be sent to MEDHA for repair.
3	Never practice any self made guideline not recommended in manual.

ANNEXURE: KAVACH – TABLE OF CONTROL

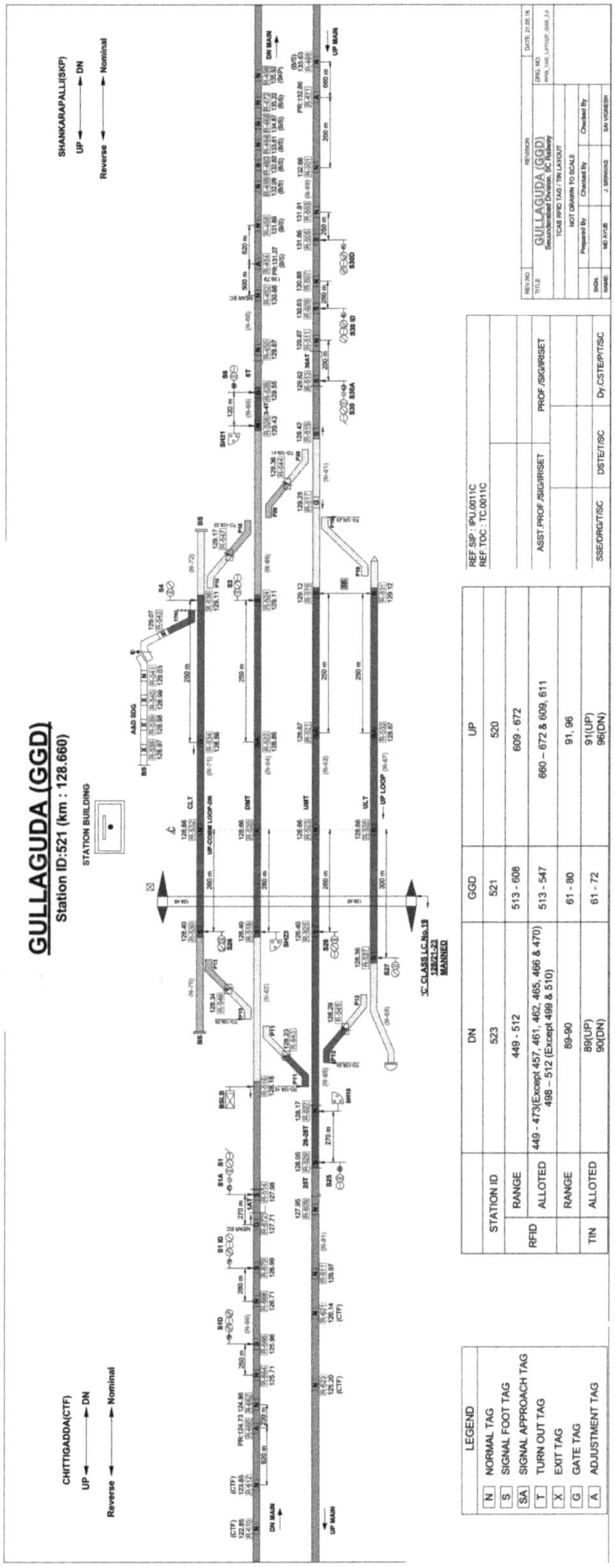

SCAN QR CODE TO ACCESS THE DIGITAL COPY OF THE STATION RFID TAG-TIN LAYOUT AND KAVACH SELECTION TABLE (PAGES 163 TO 166)

Sl. No.	Entry Signal	Exit Signal	Leading to Line	Aspect of Entry Signal	Requires Aspects of Signal	Requires Points in Route		Requires Track Circuit "UP" in Route	TINs (Track Identification Number) Requires Free	Check RFID				Turnout			Distance between Entry & Exit Signal (Meters)	Movement Authority From Foot of Entry Signal (Meters)
						Normal	Reverse			Entry Signal Foot Tag	En-Route Tags	Conflicting Route Tags	Conflicting Turnout Tags	Speed (kmph)	Distance to Commence (Meters)	Speed Restriction Distance (Meters)		
1	S1 D	S1 ID	Down Main	Green	Green	---	---	---	---	R-666	R-668	---	---	---	---	---	1030	3590
				Green	Double Yellow													3150
				Double Yellow	Green													3590
				Double Yellow	Double Yellow													3150
				Double Yellow	Yellow													2020
2	S1 ID	S1	Down Main	Green	Green	---	---	---	---	R-670	R-672	---	---	---	---	---	990	2560
				Double Yellow	Green													2560
				Double Yellow	Yellow													2120
				Double Yellow	Yellow with "Pos1"													2120
				Yellow	Green													2560
				Yellow	Red													990
3	S1	S3	Down Main	Green	Green	11,13	---	DMT	N-62, N-64	R-514	R-516,R-518,R-520,R-522	R-530,R-532,R-534	R-546	---	---	---	1130	11170
				Yellow	Yellow													1570
				Yellow	Green													11170
				Yellow	Red													1130
4	S1A	S3	Down Main	Miniature	S1 Red, S1 Blank, S3 Red	11,13	---	1AT occupied	---	R-514	R-516,R-518,R-520,R-522	R-530,R-532,R-534	R-546	15	0	1160	1130	1130
5	S1	S4	Common Loop	Yellow with "Pos1"	Red	11	13	CLT	N-62, N-70, N-71	R-514	R-516,R-546,R-530,R-532, R-534	R-518,R-520,R-522	---	30	310	850	1130	1130
					Yellow													1570
6	S1A	S4	Common Loop	Miniature	S1 Red, S1 Blank, S4 Red	11	13	1AT occupied	---	R-514	R-516,R-546,R-530,R-532, R-534	R-518,R-520,R-522	---	15	0	1160	1130	1130
7	S3	S6	Down Main	Green	Green	18,20	---	---	N-66, N-68	R-524	R-526	R-515	R-544	---	---	---	440	10040
				Yellow	Green													10040
				Yellow	Red													440
8	S4	S6	Common Loop	Yellow	Green	20	18	---	N-72, N-66, N-68	R-536	R-547,R-526	R-515	R-544	30	0	120	440	10040
				Yellow	Red													440

			GULLAGUDA (GGD) Station ID: 978 Secunderabad Division, SC Railway		REF. SIP NO : IPU.0011C TOC NO : TC.0011C & LAYOUT NO : RFID_TAG_LAYOUT_GGD_2.0
Asst Prof/ SIG/IRISET	Prof/SIG-II/IRISET		TCAS STATION SELECTION TABLE		TABLE NO: SELECTION_TABLE_GGD_1.0 PAGE 1 OF 3
				PREPARED BY	CHECKED BY
			SIGN.		
SSE/DRG/ P/T/SC	DSTE/ P/T/SC	Dy.CSTE/ P/T/SC	NAME	J.Srinivas	Sai Vignesh

Sl. No.	Entry Signal	Exit Signal	Leading to Line	Aspect of Entry Signal	Requires Aspects of Signal	Requires Points in Route		Requires Track Circuit "UP" in Route	TINs (Track Identification Number) Requires Free	Check RFID				Turnout			Distance between Entry & Exit Signal (Meters)	Movement Authority From Foot of Entry Signal (Meters)
						Normal	Reverse			Entry Signal Foot Tag	En-Route Tags	Conflicting Route Tags	Conflicting Turnout Tags	Speed (kmph)	Distance to Commence (Meters)	Speed Restriction Distance (Meters)		
9	S6	S1 D (SKP)	Down Main	Green	—	—	—	Line clear available	—	R-528	R-450,R-452	—	—	—	—	—	7570	9600
10	S30 D	S30 ID	UP Main	Green	Green	—	—	—	—	R-505	R-507	—	—	—	—	—	1030	3610
				Green	Double Yellow													3260
				Double Yellow	Green													3610
				Double Yellow	Double Yellow													3260
				Double Yellow	Yellow													2040
11	S30 ID	S30	UP Main	Green	Green	—	—	—	—	R-509	R-511	—	—	—	—	—	1010	2580
				Double Yellow	Green													2580
				Double Yellow	Yellow													2230
				Double Yellow	Yellow with "Pos4"													2230
				Double Yellow	Yellow with "Pos1"													2270
				Yellow	Green													2580
				Yellow	Red													1010
12	S30	S28	UP Main	Green	Green	20,19	—	UMT	N-61, N-63	R-513	R-515,R-517,R-519,R-521, R-523	R-536,R-542,R-541, R-534,R-532,R-524,R-522,R-520, R-518,R-531,R-533,R-535	R-544, R-547	—	—	—	1220	6770
				Yellow	Yellow													1570
				Yellow	Green													6770
				Yellow	Red													1220
13	S30A	S28	UP Main	Miniature	S30 Red, S30 Blank, S28 Red	20,19	—	—	30AT occupied	R-513	R-515,R-517,R-519,R-521, R-523	R-536,R-542,R-541, R-534,R-532,R-524,R-522,R-520, R-518,R-531,R-533,R-535	R-544, R-547	15	0	1250	1220	1220
14	S30	S27	UP Loop	Yellow with "Pos1"	Red	20	19	ULT	N-61, N-67	R-513	R-515,R-517,R-531,R-533, R-535	R-536,R-542,R-541, R-534,R-532,R-524,R-522,R-520, R-518,R-519,R-521,R-523	R-544, R-547	30	390	900	1260	1260
					Yellow													1570
15	S30A	S27	UP Loop	Miniature	S30 Red, S30 Blank, S27 Red	20	19	—	30AT occupied	R-513	R-515,R-517,R-531,R-533, R-535	R-536,R-542,R-541, R-534,R-532,R-524,R-522,R-520, R-518,R-519,R-521,R-523	R-544, R-547	15	0	1290	1260	1260

			GULLAGUDA (GGD) Station ID: 978 Secunderabad Division, SC Railway			REF. SIP NO : IPU.0011C TOC NO : TC.0011C & LAYOUT NO : RFID_TAG_LAYOUT_GGD_2.0
Asst Prof/ SIG/IRISET	Prof/SIG-I/IRISET		TCAS STATION SELECTION TABLE			TABLE NO: SELECTION_TABLE_GGD_1.0 PAGE 2 OF 3
				PREPARED BY	CHECKED BY	
			SIGN.			
SSE/DRG/ P/T/SC	DSTE/ P/T/SC	Dy.CSTE/ P/T/SC	NAME	J.Srinivas	Sai Vignesh	

Sl. No.	Entry Signal	Exit Signal	Leading to Line	Aspect of Entry Signal	Requires Aspects of Signal	Requires Points in Route		Requires Track Circuit "UP" in Route	TINs (Track Identification Number) Requires Free	Check RFID				Turnout			Distance between Entry & Exit Signal (Meters)	Movement Authority From Foot of Entry Signal (Meters)
						Normal	Reverse			Entry Signal Foot Tag	En-Route Tags	Conflicting Route Tags	Conflicting Turnout Tags	Speed (kmph)	Distance to Commence (Meters)	Speed Restriction Distance (Meters)		
16	S30	S26	Common Loop	Yellow with "Pos4"	Red	17KL	20,18	CLT	N-61, N-66, N-72, N-71	R-513	R-515,R-544,R-547,R-536, R-534,R-532	R-542,R-541,R-524,R-522,R-520,R-518, R-517,R-519,R-521,R-523, R-531,R-533,R-535	—	15	210	1040	1220	1220
					Yellow													1570
17	S30A	S26	Common Loop	Miniature	S30 Red, S30 Blank, S26 Red	17KL	20,18	—	30AT occupied	R-513	R-515,R-544,R-547,R-536, R-534,R-532	R-542,R-541,R-524,R-522,R-520,R-518, R-517,R-519,R-521,R-523, R-531,R-533,R-535	—	15	0	1250	1220	1220
18	S28	S25	UP Main	Green	Green	12,11	—	—	N-65	R-525	R--527	—	-	—	—	—	350	5550
				Yellow	Green													5550
				Yellow	Red													350
19	S27	S25	UP Loop	Yellow	Green	11	12	—	N69, N-65	R-537	R-545,R-527	—	-	30	0	130	310	5510
				Yellow	Red													310
20	S26	S25	Common Loop	Yellow	Green	—	13,11	—	N-70, N-62, N-65	R-530	R-546,R-543,R-527	R-516	-	15	0	230	350	5550
				Yellow	Red													350
21	S25	S1 D (CTF)	UP Main	Green	—	—	—	Line clear available	—	R-529	R-609,R-611	—	-	—	—	—	3000	5200

Action for generating SOS to prevent HOC on received communication for Block Section TIN		
Entry Signal for Block Selection TIN	RFID Tags	TIN
S6	R-450,R-452,R-454,R-456, R-458,R-460,R-464,R-468, R-472,R-498,R-500,R-502, R-504,R-506,R-508,R-512, R-354	90
S25	R-609,R-611,R-621,R-623, R-673,R-675,R-677,R-679, R-681	91

Last Signal of Station	Movement Authority
S6	9600
S25	5200

Rev No.	Revision	Date
1.0	Initial version	24-11-2021

		GULLAGUDA (GGD) Station ID: 978 Secunderabad Division, SC Railway			REF. SIP NO : IPU.0011C TOC NO : TC.0011C & LAYOUT NO : RFID_TAG_LAYOUT_GGD_2.0
Asst Prof/ SIG/IRISET	Prof/SIG-I/IRISET	TCAS STATION SELECTION TABLE			TABLE NO: SELECTION_TABLE_GGD_1.0 PAGE 3 OF 3
			PREPARED BY	CHECKED BY	
		SIGN.			
SSE/DRG/ P/T/SC	DSTE/ P/T/SC	Dy.CSTE/ P/T/SC	NAME	J.Srinivas	Sai Vignesh

REVIEW QUESTIONS

CHAPTER - 1

SUBJECTIVE

1. List out the equipments of track side Kavach and Loco Kavach.
2. Write the features of Driver Machine Interface (DMI).
3. Explain in brief working of Kavach.

OBJECTIVE

1. Expand the following terms used in Kavach:
   (a) MA :
   (b) PSR :
   (c) SPAD :
   (d) NMS :
   (e) KMS :
   (f) SB :
   (g) EB :
   (h) FSB :

CHAPTER - 2

SUBJECTIVE

1. What is the purpose of RF TAG
2. How many types of RF TAGs and explain them
3. What are the Information programmed to RF TAG

OBJECTIVE

1. RF TAG is ______ type of device

   (a) Active (b) Passive (c) a & b (d) None

2. All RF TAGs are placed in the ______ architecture

   (a) 2 out of 2 (b) 2 out of 3 (c) a & b (d) None

3. Distance between normal tags incase of block section

   (a) 1 KM (b) 5 KM (c) 3.5 KM (d) None

CHAPTER - 3

SUBJECTIVE

1. Explain radio modem functional requirements
2. Explain about TDMA/FDMA, Timeslots
3. Explain radio modem requirements
4. Describes the LTE Radio modem requirements to be used for the purpose of Kavach System.

OBJECTIVE

1. Communication packets exchanged between Station Kavach and Loco Kavach shall be of a maximum size of ______ bits.

   (a) 1024 (b) 512 (c) 4096 (d) 10000

2. Communication between the Stationary Kavach and Loco Kavach shall be Over-The- Air using Multiple Access & each Multiple Access frame cycle shall be of _____ milli seconds

   (a) 2000 (b) 3000 (c) 4000 (d) 5000

3. Loco Kavach shall use transmit frequency (f_0) in block section and at the times of emergency situations

   (a) SoS (b) head-on collisions
   (c) Rear end collision (d) All

4. Time slots P28, P29, P30, P31, P32, P33, P67, P68, P69, P70, P71 and P72 shall be used by the Loco Kavach for broad casting communication packets in the ______ section

   (a) Block (b) Station (c) a & b (d) None

5. LTE modems working voltages is ________

   (a) 24 V DC (b) 110 V AC (c) 60 V AC (d) a & b

CHAPTER - 4

SUBJECTIVE

1. Describe about KMS with block diagram
2. Explain the following techniques used for Kavach communications
   (a) Radio Communication
   (b) GSM /GPRS Communication
   (c) GPS/GNSS Communication
3. Explain communication mandatory zone and Communication in overlap zone
4. List out the fault messages logged by Stationary Kavach
5. List out the fault messages logged by Loco Kavach
6. Explain about Authentication Key (K_A) and Session Key (K_S)

CHAPTER - 5

SUBJECTIVE

1. Explain the features of Kavach Network Monitoting System (NMS)?
2. Explain Loco Kavach communication with NMS and what information is transmitted from Loco Kavach to NMS?
3. Explain what information is transmitted from Station Kavach to NMS?

OBJECTIVE

1. Station Kavach to NMS connectivity, how many Number of Station Kavach units connected in one E1 Ring? ( )
   (a) 5-7 (b) 7-9 (c) 10-16 (d) 20-25
2. Station Kavach communication with NMS is connected through ______ interface? ( )
   (a) E1 (b) GPRS (c) Serial (d) Radio
3. Loco Kavach communication with NMS is connected through ________ interface? ( )
   (a) E1 (b) GPRS (c) GSM (d) Radio
4. Station Kavach connected to adjacent station Kavach through _______ interface for connecting to NMS?
   (a) E1 (b) GPRS (c) GSM (d) Radio

5. In NMS, IB huts are_____________
   (a) Directly connected to NMS (b) Connected to Nearest Station Kavach
   (c) Not Connected (d) Not required to connect

CHAPTER - 6

SUBJECTIVE

1. How RIU connected to the station Kavach
2. Explain about Power supply arrangement of RIU
3. Explain about any OEM RIU architecture

OBJECTIVE

1. Station Kavach can be connecting minimum_____ of RIU in one direction.
   (a) 6 (b) 5 (c) a & b (d) None
2. RIU of Kavach unit consist of vital input models with minimum_____ architecture
   (a) 2 out of 2 (b) 2 out of 3 (c) a & b (d) None
3. RIU is provided at _____________ places
   (a) Automatic section (b) LC gate
   (c) IB (d) a, b & c
4. How many km can RIU may be placed at all locations from the station ____
   (a) 3 (b) 5 (c) 8 (d) None
5. Working voltage of RIU is ____________
   (a) 230 V AC (b) 500 V AC
   (c) 12 V DC (d) 60 V DC

Terms and Definitions

1. **Automatic Train Protection (ATP):** Automatic Train Protection (ATP) is a type of train protection system which continually checks that the speed of a train is compatible with the permitted speed allowed by signalling, including automatic stop at certain signal aspects. If it is not, ATP activates an emergency brake to stop the train. In other words it provides Fail safe protection against over speed, collision & other hazardous conditions through train detection, train separation & interlocking. The main functions of ATP are:
   - Detection and Prevention of SPAD
   - Display of signal aspect, movement authority, target distance and speed.
   - Continuous train control.
   - Protection for Permanent and temporary speed restriction.
2. **Berthing Track:** This is the designated section of the track in station section on which trains normally stop e.g. Platform lines etc.
3. **Block Section:** The portion of the running line between two block stations onto which no running train may enter until Line Clear has been received from the block station at the other end of the block section.
4. **Communication Mandatory Zone:** It is the area on track where continuous communication between Loco Kavach and Stationary Kavach is required.
5. **Direction of Movement of Loco:** This is the direction of the train as per Loco cab control e.g. Forward or Reverse or Neutral.
6. **Dynamic Speed Profile:** The speed-distance curve which a train shall follow without violating the static train speed profile till the end of movement authority. This curve depends on the braking characteristics of the train and the train length.

7. **Emergency Brake:** It is fail-safe, open-loop braking to a complete stop, maximum stopping distance is assured, brake is irreversible. It involves shutting off power and full application of brakes without any loss of time.
8. **Emergency Braking Distance:** Emergency braking distance is the distance travelled by train before coming to a stop by sudden application of brake at one stretch.
9. **End of Authority (EOA):** Location up to which the train is permitted to proceed and where target speed is zero.
10. **Factory Acceptance Tests:** Tests carried out by installing some equipment in the lab to prove that the system performs in accordance with this specification & the application data.
11. **Interlocking:** An arrangement of signals, points and other appliances, operated from a panel or lever frame, so interconnected by mechanical locking or electrical locking a fixed stop signal in IBS or both that their operation must take place in proper sequence to ensure safety.
12. **Movement Authority:** The distance upto which the train is permitted to travel without danger.
13. **On-board Equipment:** This subsystem consists of a combination of vital and non-vital equipment located on the passenger train-sets and maintenance vehicles. Vital equipment is used to fulfil the ATP functions; non vital equipment is used to fulfil all non ATP functions such as ATO and displays. The equipment includes processors, firmware, software and electronics, operator displays, operator panel, data radios and antennas, transponder/ balise antennas, code pick-up antennas, network components, GPS receiver and antennas, tachometers and other sensors.
14. **Permanent Speed Restriction (PSR):** For various reasons, although mainly because of track geometry (curvature, etc.), it is necessary to limit the speed at which trains may travel over certain sections of the railway. These places are subject to what are termed 'permanent speed restrictions' (PSRs). In some instances, different speeds are specified for specific types of trains.
15. **Service Brake:** Service brake is a non-emergency brake application–which is reversible. It involves only the shutting off the power and the gradual application of brakes.
16. **Service Braking Distance:** It is the distance required to stop the train running at the maximum permissible speed of the line, at such a rate of deceleration that the passengers do not suffer discomfort or alarm.
17. **SPAD:** SPAD stands for 'Signal Passed at Danger' and occurs when a train passes a signal in the 'on' (Red) position without authority.
18. **Static Speed Profile:** The Static Speed Profile (SSP) is a description of the fixed speed restrictions at a resolution of 5 Kmph for a part of track sent from trackside to train.
19. **Station Section:** The portion of station limits which can be used for shunting even after granting Line clear to station in Rear.

20. **Tachometer:** A tachometer is an instrument measuring the rotation speed of a shaft or disk, as in a motor or other machine. The device usually displays the revolutions per minute (RPM) on a calibrated analogue dial or digital display.
21. **Temporary Speed Restriction (TSR):** The object of a TSR is to reduce the speed of Rail Traffic to ensure safe passage over a Section of Track when the Track is not safe for Normal Speed. A TSR is applied by a Maintenance Representative of concerned department such as Engg. or S&T. A TSR overrides any existing higher speed.
22. **RFID Tags:** RFID stands for Radio Frequency Identification. An RFID tag works by transmitting and receiving information via an antenna and a microchip. RFID tags are fitted on track in station section, point zones, near Signals & track in block section for giving Trackside information to Loco Kavach unit.
23. **Radio Hole:** Strong fading of the radio signal at some position in space along an air to air or air to ground path; the effect is caused by the abnormal refraction of radio waves.

Index

www.ingramcontent.com/pod-product-compliance
Ingram Content Group UK Ltd.
Pitfield, Milton Keynes, MK11 3LW, UK
UKHW061533190726
13845UKWH00005B/83

9 789395 038355